Information: A Very Short Introduction

VERY SHORT INTRODUCTIONS are for anyone wanting a stimulating and accessible way in to a new subject. They are written by experts and have been translated into more than 40 different languages. The series began in 1995 and now covers a wide variety of topics in every discipline. The VSI library contains nearly 400 volumes—a Very Short Introduction to everything from Indian philosophy to psychology and American history—and continues to grow in every subject area.

Very Short Introductions available now:

ACCOUNTING Christopher Nobes
ADVERTISING Winston Fletcher
AFRICAN HISTORY John Parker and Richard Rathbone
AGNOSTICISM Robin Le Poidevin
ALEXANDER THE GREAT Hugh Bowden
AMERICAN HISTORY Paul S. Boyer
AMERICAN IMMIGRATION David A. Gerber
AMERICAN POLITICAL PARTIES AND ELECTIONS L. Sandy Maisel
AMERICAN POLITICS Richard M. Valelly
THE AMERICAN PRESIDENCY Charles O. Jones
ANAESTHESIA Aidan O'Donnell
ANARCHISM Colin Ward
ANCIENT EGYPT Ian Shaw
ANCIENT GREECE Paul Cartledge
THE ANCIENT NEAR EAST Amanda H. Podany
ANCIENT PHILOSOPHY Julia Annas
ANCIENT WARFARE Harry Sidebottom
ANGELS David Albert Jones
ANGLICANISM Mark Chapman
THE ANGLO-SAXON AGE John Blair
THE ANIMAL KINGDOM Peter Holland
ANIMAL RIGHTS David DeGrazia
THE ANTARCTIC Klaus Dodds
ANTISEMITISM Steven Beller
ANXIETY Daniel Freeman and Jason Freeman
THE APOCRYPHAL GOSPELS Paul Foster
ARCHAEOLOGY Paul Bahn
ARCHITECTURE Andrew Ballantyne
ARISTOCRACY William Doyle
ARISTOTLE Jonathan Barnes
ART HISTORY Dana Arnold
ART THEORY Cynthia Freeland
ASTROBIOLOGY David C. Catling
ATHEISM Julian Baggini
AUGUSTINE Henry Chadwick
AUSTRALIA Kenneth Morgan
AUTISM Uta Frith
THE AVANT GARDE David Cottington
THE AZTECS Davíd Carrasco
BACTERIA Sebastian G. B. Amyes
BARTHES Jonathan Culler
THE BEATS David Sterritt
BEAUTY Roger Scruton
BESTSELLERS John Sutherland
THE BIBLE John Riches
BIBLICAL ARCHAEOLOGY Eric H. Cline
BIOGRAPHY Hermione Lee
THE BLUES Elijah Wald
THE BOOK OF MORMON Terryl Givens
BORDERS Alexander C. Diener and Joshua Hagen
THE BRAIN Michael O'Shea
THE BRITISH CONSTITUTION Martin Loughlin
THE BRITISH EMPIRE Ashley Jackson
BRITISH POLITICS Anthony Wright
BUDDHA Michael Carrithers
BUDDHISM Damien Keown
BUDDHIST ETHICS Damien Keown
CANCER Nicholas James
CAPITALISM James Fulcher
CATHOLICISM Gerald O'Collins
CAUSATION Stephen Mumford and Rani Lill Anjum
THE CELL Terence Allenand Graham Cowling
THE CELTS Barry Cunliffe
CHAOS Leonard Smith
CHILDREN'S LITERATURE Kimberley Reynolds
CHINESE LITERATURE Sabina Knight
CHOICE THEORY Michael Allingham
CHRISTIAN ART Beth Williamson
CHRISTIAN ETHICS D. Stephen Long
CHRISTIANITY Linda Woodhead
CITIZENSHIP Richard Bellamy
CIVIL ENGINEERING David Muir Wood
CLASSICAL LITERATURE William Allan
CLASSICAL MYTHOLOGY Helen Morales
CLASSICS Mary Beard and John Henderson
CLAUSEWITZ Michael Howard
CLIMATE Mark Maslin
THE COLD WAR Robert McMahon
COLONIAL AMERICA Alan Taylor

COLONIAL LATIN AMERICAN LITERATURE Rolena Adorno
COMEDY Matthew Bevis
COMMUNISM Leslie Holmes
COMPLEXITY John H. Holland
THE COMPUTER Darrel Ince
THE CONQUISTADORS Matthew Restall and Felipe Fernández-Armesto
CONSCIENCE Paul Strohm
CONSCIOUSNESS Susan Blackmore
CONTEMPORARY ART Julian Stallabrass
CONTEMPORARY FICTION Robert Eaglestone
CONTINENTAL PHILOSOPHY Simon Critchley
CORAL REEFS Charles Sheppard
COSMOLOGY Peter Coles
CRITICAL THEORY Stephen Eric Bronner
THE CRUSADES Christopher Tyerman
CRYPTOGRAPHY Fred Piper and Sean Murphy
THE CULTURAL REVOLUTION Richard Curt Kraus
DADA AND SURREALISM David Hopkins
DARWIN Jonathan Howard
THE DEAD SEA SCROLLS Timothy Lim
DEMOCRACY Bernard Crick
DERRIDA Simon Glendinning
DESCARTES Tom Sorell
DESERTS Nick Middleton
DESIGN John Heskett
DEVELOPMENTAL BIOLOGY Lewis Wolpert
THE DEVIL Darren Oldridge
DIASPORA Kevin Kenny
DICTIONARIES Lynda Mugglestone
DINOSAURS David Norman
DIPLOMACY Joseph M. Siracusa
DOCUMENTARY FILM Patricia Aufderheide
DREAMING J. Allan Hobson
DRUGS Leslie Iversen
DRUIDS Barry Cunliffe
EARLY MUSIC Thomas Forrest Kelly
THE EARTH Martin Redfern
ECONOMICS Partha Dasgupta
EDUCATION Gary Thomas
EGYPTIAN MYTH Geraldine Pinch
EIGHTEENTH-CENTURY BRITAIN Paul Langford
THE ELEMENTS Philip Ball
EMOTION Dylan Evans
EMPIRE Stephen Howe
ENGELS Terrell Carver
ENGINEERING David Blockley
ENGLISH LITERATURE Jonathan Bate
ENVIRONMENTAL ECONOMICS Stephen Smith
EPIDEMIOLOGY Rodolfo Saracci
ETHICS Simon Blackburn
THE EUROPEAN UNION John Pinder and Simon Usherwood
EVOLUTION Brian and Deborah Charlesworth
EXISTENTIALISM Thomas Flynn
THE EYE Michael Land
FAMILY LAW Jonathan Herring
FASCISM Kevin Passmore
FASHION Rebecca Arnold
FEMINISM Margaret Walters
FILM Michael Wood
FILM MUSIC Kathryn Kalinak
THE FIRST WORLD WAR Michael Howard
FOLK MUSIC Mark Slobin
FOOD John Krebs
FORENSIC PSYCHOLOGY David Canter
FORENSIC SCIENCE Jim Fraser
FOSSILS Keith Thomson
FOUCAULT Gary Gutting
FRACTALS Kenneth Falconer
FREE SPEECH Nigel Warburton
FREE WILL Thomas Pink
FRENCH LITERATURE John D. Lyons
THE FRENCH REVOLUTION William Doyle
FREUD Anthony Storr
FUNDAMENTALISM Malise Ruthven
GALAXIES John Gribbin
GALILEO Stillman Drake
GAME THEORY Ken Binmore
GANDHI Bhikhu Parekh
GENIUS Andrew Robinson
GEOGRAPHY John Matthews and David Herbert
GEOPOLITICS Klaus Dodds
GERMAN LITERATURE Nicholas Boyle
GERMAN PHILOSOPHY Andrew Bowie
GLOBAL CATASTROPHES Bill McGuire
GLOBAL ECONOMIC HISTORY Robert C. Allen
GLOBAL WARMING Mark Maslin
GLOBALIZATION Manfred Steger
THE GOTHIC Nick Groom
GOVERNANCE Mark Bevir
THE GREAT DEPRESSION AND THE NEW DEAL Eric Rauchway
HABERMAS James Gordon Finlayson
HAPPINESS Daniel M. Haybron
HEGEL Peter Singer
HEIDEGGER Michael Inwood
HERODOTUS Jennifer T. Roberts
HIEROGLYPHS Penelope Wilson
HINDUISM Kim Knott
HISTORY John H. Arnold
THE HISTORY OF ASTRONOMY Michael Hoskin
THE HISTORY OF LIFE Michael Benton
THE HISTORY OF MATHEMATICS Jacqueline Stedall
THE HISTORY OF MEDICINE William Bynum
THE HISTORY OF TIME Leofranc Holford-Strevens

HIV/AIDS Alan Whiteside
HOBBES Richard Tuck
HORMONES Martin Luck
HUMAN EVOLUTION Bernard Wood
HUMAN RIGHTS Andrew Clapham
HUMANISM Stephen Law
HUME A. J. Ayer
HUMOUR Noël Carroll
THE ICE AGE Jamie Woodward
IDEOLOGY Michael Freeden
INDIAN PHILOSOPHY Sue Hamilton
INFORMATION Luciano Floridi
INNOVATION Mark Dodgson and David Gann
INTELLIGENCE Ian J. Deary
INTERNATIONAL MIGRATION Khalid Koser
INTERNATIONAL RELATIONS Paul Wilkinson
INTERNATIONAL SECURITY
Christopher S. Browning
ISLAM Malise Ruthven
ISLAMIC HISTORY Adam Silverstein
ITALIAN LITERATURE
Peter Hainsworth and David Robey
JESUS Richard Bauckham
JOURNALISM Ian Hargreaves
JUDAISM Norman Solomon
JUNG Anthony Stevens
KABBALAH Joseph Dan
KAFKA Ritchie Robertson
KANT Roger Scruton
KEYNES Robert Skidelsky
KIERKEGAARD Patrick Gardiner
THE KORAN Michael Cook
LANDSCAPE ARCHITECTURE
Ian H. Thompson
LANDSCAPES AND GEOMORPHOLOGY
Andrew Goudie and Heather Viles
LANGUAGES Stephen R. Anderson
LATE ANTIQUITY Gillian Clark
LAW Raymond Wacks
THE LAWS OF THERMODYNAMICS
Peter Atkins
LEADERSHIP Keith Grint
LINCOLN Allen C. Guelzo
LINGUISTICS Peter Matthews
LITERARY THEORY Jonathan Culler
LOCKE John Dunn
LOGIC Graham Priest
MACHIAVELLI Quentin Skinner
MADNESS Andrew Scull
MAGIC Owen Davies
MAGNA CARTA Nicholas Vincent
MAGNETISM Stephen Blundell
MALTHUS Donald Winch
MANAGEMENT John Hendry
MAO Delia Davin
MARINE BIOLOGY Philip V. Mladenov
THE MARQUIS DE SADE John Phillips
MARTIN LUTHER Scott H. Hendrix
MARTYRDOM Jolyon Mitchell
MARX Peter Singer
MATHEMATICS Timothy Gowers
THE MEANING OF LIFE Terry Eagleton
MEDICAL ETHICS Tony Hope
MEDICAL LAW Charles Foster
MEDIEVAL BRITAIN
John Gillingham and Ralph A. Griffiths
MEMORY Jonathan K. Foster
METAPHYSICS Stephen Mumford
MICHAEL FARADAY Frank A.J.L. James
MICROECONOMICS Avinash Dixit
MODERN ART David Cottington
MODERN CHINA Rana Mitter
MODERN FRANCE Vanessa R. Schwartz
MODERN IRELAND Senia Pašeta
MODERN JAPAN Christopher Goto-Jones
MODERN LATIN AMERICAN LITERATURE
Roberto González Echevarría
MODERN WAR Richard English
MODERNISM Christopher Butler
MOLECULES Philip Ball
THE MONGOLS Morris Rossabi
MORMONISM Richard Lyman Bushman
MUHAMMAD Jonathan A.C. Brown
MULTICULTURALISM Ali Rattansi
MUSIC Nicholas Cook
MYTH Robert A. Segal
THE NAPOLEONIC WARS Mike Rapport
NATIONALISM Steven Grosby
NELSON MANDELA Elleke Boehmer
NEOLIBERALISM Manfred Steger and Ravi Roy
NETWORKS
Guido Caldarelli and Michele Catanzaro
THE NEW TESTAMENT Luke Timothy Johnson
THE NEW TESTAMENT AS LITERATURE
Kyle Keefer
NEWTON Robert Iliffe
NIETZSCHE Michael Tanner
NINETEENTH-CENTURY BRITAIN
Christopher Harvie and H. C. G. Matthew
THE NORMAN CONQUEST George Garnett
NORTH AMERICAN INDIANS
Theda Perdue and Michael D. Green
NORTHERN IRELAND Marc Mulholland
NOTHING Frank Close
NUCLEAR POWER Maxwell Irvine
NUCLEAR WEAPONS Joseph M. Siracusa
NUMBERS Peter M. Higgins
NUTRITION David A. Bender
OBJECTIVITY Stephen Gaukroger
THE OLD TESTAMENT Michael D. Coogan
THE ORCHESTRA D. Kern Holoman
ORGANIZATIONS Mary Jo Hatch
PAGANISM Owen Davies
THE PALESTINIAN-ISRAELI CONFLICT
Martin Bunton
PARTICLE PHYSICS Frank Close

PAUL E. P. Sanders
PENTECOSTALISM William K. Kay
THE PERIODIC TABLE Eric R. Scerri
PHILOSOPHY Edward Craig
PHILOSOPHY OF LAW Raymond Wacks
PHILOSOPHY OF SCIENCE Samir Okasha
PHOTOGRAPHY Steve Edwards
PLAGUE Paul Slack
PLANETS David A. Rothery
PLANTS Timothy Walker
PLATO Julia Annas
POLITICAL PHILOSOPHY David Miller
POLITICS Kenneth Minogue
POSTCOLONIALISM Robert Young
POSTMODERNISM Christopher Butler
POSTSTRUCTURALISM Catherine Belsey
PREHISTORY Chris Gosden
PRESOCRATIC PHILOSOPHY
Catherine Osborne
PRIVACY Raymond Wacks
PROBABILITY John Haigh
PROGRESSIVISM Walter Nugent
PROTESTANTISM Mark A. Noll
PSYCHIATRY Tom Burns
PSYCHOLOGY Gillian Butler and Freda McManus
PURITANISM Francis J. Bremer
THE QUAKERS Pink Dandelion
QUANTUM THEORY John Polkinghorne
RACISM Ali Rattansi
RADIOACTIVITY Claudio Tuniz
RASTAFARI Ennis B. Edmonds
THE REAGAN REVOLUTION Gil Troy
REALITY Jan Westerhoff
THE REFORMATION Peter Marshall
RELATIVITY Russell Stannard
RELIGION IN AMERICA Timothy Beal
THE RENAISSANCE Jerry Brotton
RENAISSANCE ART Geraldine A. Johnson
REVOLUTIONS Jack A. Goldstone
RHETORIC Richard Toye
RISK Baruch Fischhoff and John Kadvany
RIVERS Nick Middleton
ROBOTICS Alan Winfield
ROMAN BRITAIN Peter Salway
THE ROMAN EMPIRE Christopher Kelly
THE ROMAN REPUBLIC David M. Gwynn
ROMANTICISM Michael Ferber
ROUSSEAU Robert Wokler
RUSSELL A. C. Grayling
RUSSIAN HISTORY Geoffrey Hosking
RUSSIAN LITERATURE Catriona Kelly
THE RUSSIAN REVOLUTION S. A. Smith
SCHIZOPHRENIA
Chris Frith and Eve Johnstone
SCHOPENHAUER Christopher Janaway
SCIENCE AND RELIGION Thomas Dixon
SCIENCE FICTION David Seed
THE SCIENTIFIC REVOLUTION
Lawrence M. Principe
SCOTLAND Rab Houston
SEXUALITY Véronique Mottier
SHAKESPEARE Germaine Greer
SIKHISM Eleanor Nesbitt
THE SILK ROAD James A. Millward
SLEEP Steven W. Lockley and Russell G. Foster
SOCIAL AND CULTURAL
ANTHROPOLOGY
John Monaghan and Peter Just
SOCIALISM Michael Newman
SOCIOLINGUISTICS John Edwards
SOCIOLOGY Steve Bruce
SOCRATES C. C. W. Taylor
THE SOVIET UNION Stephen Lovell
THE SPANISH CIVIL WAR Helen Graham
SPANISH LITERATURE Jo Labanyi
SPINOZA Roger Scruton
SPIRITUALITY Philip Sheldrake
STARS Andrew King
STATISTICS David J. Hand
STEM CELLS Jonathan Slack
STUART BRITAIN John Morrill
SUPERCONDUCTIVITY Stephen Blundell
SYMMETRY Ian Stewart
TEETH Peter S. Ungar
TERRORISM Charles Townshend
THEOLOGY David F. Ford
THOMAS AQUINAS Fergus Kerr
THOUGHT Tim Bayne
TIBETAN BUDDHISM Matthew T. Kapstein
TOCQUEVILLE Harvey C. Mansfield
TRAGEDY Adrian Poole
THE TROJAN WAR Eric H. Cline
TRUST Katherine Hawley
THE TUDORS John Guy
TWENTIETH-CENTURY BRITAIN
Kenneth O. Morgan
THE UNITED NATIONS Jussi M. Hanhimäki
THE U.S. CONGRESS Donald A. Ritchie
THE U.S. SUPREME COURT
Linda Greenhouse
UTOPIANISM Lyman Tower Sargent
THE VIKINGS Julian Richards
VIRUSES Dorothy H. Crawford
WITCHCRAFT Malcolm Gaskill
WITTGENSTEIN A. C. Grayling
WORK Stephen Fineman
WORLD MUSIC Philip Bohlman
THE WORLD TRADE ORGANIZATION
Amrita Narlikar
WRITING AND SCRIPT Andrew Robinson

Luciano Floridi

# INFORMATION

## A Very Short Introduction

Great Clarendon Street, Oxford OX2 6DP

Oxford University Press is a department of the University of Oxford.
It furthers the University's objective of excellence in research, scholarship,
and education by publishing worldwide in

Oxford New York

Auckland Cape Town Dar es Salaam Hong Kong Karachi
Kuala Lumpur Madrid Melbourne Mexico City Nairobi
New Delhi Shanghai Taipei Toronto

With offices in

Argentina Austria Brazil Chile Czech Republic France Greece
Guatemala Hungary Italy Japan Poland Portugal Singapore
South Korea Switzerland Thailand Turkey Ukraine Vietnam



Published in the United States
by Oxford University Press Inc., New York



First published 2010



British Library Cataloguing in Publication Data
Data available

Library of Congress Cataloging in Publication Data
Data available

Typeset by SPI Publisher Services, Pondicherry, India
Printed in Great Britain by
Ashford Colour Press Ltd, Gosport, Hampshire

ISBN 978-0-19-955137-8

7 9 10 8 6

# Contents

# Contents

# Acknowledgements

Among the many people who helped me in writing this book, I would like to thank explicitly Kerstin Demata, Emma Marchant, and Latha Menon, all at Oxford University Press, for their encouragement, support, editorial input, and ample but luckily finite patience, thanks to which I managed to finish the book; Robert Schaback, for his kind help while I was working at the Institut für Numerische und Angewandte Mathematik of the Georg-August-Universität in Göttingen and his comments on the penultimate version of the manuscript; David Davenport, Ugo Pagallo, and Christoph Schulz for their feedback; and my wife Anna Christina De Ozorio Nobre, for her invaluable input and comments, especially on Chapter 6, and for having made our life unbelievably joyful. I am very grateful to the Akademie der Wissenschaften in Göttingen, for the privilege of being elected Gauss Professor during the academic year 2008–9, and to the University of Hertfordshire, for having been generous with my teaching schedule while visiting Göttingen and completing this book.

# List of illustrations

# List of tables

# Introduction

The goal of this volume is to provide an outline of what information is, of its manifold nature, of the roles that it plays in several scientific contexts, and of the social and ethical issues raised by its growing importance. The outline is necessarily selective, or it would be neither very short nor introductory. My hope is that it will help the reader to make sense of the large variety of informational phenomena with which we deal on a daily basis, of their profound and fundamental importance, and hence of the information society in which we live.

Information is notorious for coming in many forms and having many meanings. It can be associated with several explanations, depending on the perspective adopted and the requirements and desiderata one has in mind. The father of information theory, Claude Shannon (1916–2001), for one, was very cautious:

> The word 'information' has been given different meanings by various writers in the general field of information theory. It is likely that at least a number of these will prove sufficiently useful in certain applications to deserve further study and permanent recognition. *It is hardly to be expected that a single concept of information would satisfactorily account for the numerous possible applications of this general field.* (italics added)

Indeed, Warren Weaver (1894–1978), one of the pioneers of machine translation and co-author with Shannon of *The Mathematical Theory of Communication*, supported a tripartite analysis of information in terms of

1) technical problems concerning the quantification of information and dealt with by Shannon's theory;
2) semantic problems relating to meaning and truth; and
3) what he called 'influential' problems concerning the impact and effectiveness of information on human behaviour, which he thought had to play an equally important role.

Shannon and Weaver provide two early examples of the problems raised by any analysis of information. The plethora of different interpretations can be confusing, and complaints about misunderstandings and misuses of the very idea of information are frequently expressed, even if apparently to no avail. This book seeks to provide a map of the main senses in which one may speak of information. The map is drawn by relying on an initial account of information based on the concept of data. Unfortunately, even such a minimalist account is open to disagreement. In favour of this approach, one may say that at least it is much less controversial than others. Of course, a conceptual analysis must start somewhere. This often means adopting some working definition of the object under scrutiny. But it is not this commonplace that I wish to emphasize here. The difficulty is rather more daunting. Work on the concept of information is still at that lamentable stage when disagreement affects even the way in which the problems themselves are provisionally phrased and framed. So the various 'you are here' signals in this book might be placed elsewhere. The whole purpose is to put the family of concepts of information firmly on the map and thus make possible further adjustments and re-orientations.

# Chapter 1
# The information revolution

## The emergence of the information society

History has many metrics. Some are natural and circular, relying on recurring seasons and planetary motions. Some are social or political and linear, being determined, for example, by the succession of Olympic Games, or the number of years since the founding of the city of Rome (*ab urbe condita*), or the ascension of a king. Still others are religious and have a V-shape, counting years before and after a particular event, such as the birth of Christ. There are larger periods that encompass smaller ones, named after influential styles (Baroque), people (Victorian era), particular circumstances (Cold War), or some new technology (nuclear age). What all these and many other metrics have in common is that they are all *historical*, in the strict sense that they all depend on the development of systems to record events and hence accumulate and transmit information about the past. No records, no history, so history is actually synonymous with the information age, since *prehistory* is that age in human development that precedes the availability of recording systems.

It follows that one may reasonably argue that humanity has been living in various kinds of information societies at least since the Bronze Age, the era that marks the invention of writing in Mesopotamia and other regions of the world (4th millennium BC).

And yet, this is not what is typically meant by the information revolution. There may be many explanations, but one seems more convincing than any other: only very recently has human progress and welfare begun to depend mostly on the successful and efficient management of the life cycle of information.

The life cycle of information typically includes the following phases: *occurrence* (discovering, designing, authoring, etc.), *transmission* (networking, distributing, accessing, retrieving, transmitting, etc.), *processing* and *management* (collecting, validating, modifying, organizing, indexing, classifying, filtering, updating, sorting, storing, etc.), and *usage* (monitoring, modelling, analysing, explaining, planning, forecasting, decision-making, instructing, educating, learning, etc.). Figure 1 provides a simplified illustration.



Now, imagine Figure 1 to be like a clock. The length of time that the evolution of information life cycles has taken to bring about the information society should not be surprising. According to recent estimates, life on Earth will last for another billion years, until it will be destroyed by the increase in solar temperature. So imagine an historian writing in the near future, say in a million years. She may consider it normal, and perhaps even elegantly symmetrical, that it took roughly six millennia for the agricultural revolution to produce its full effect, from its beginning in the Neolithic (10th millennium BC), until the Bronze Age, and then another six millennia for the information revolution to bear its main fruit, from the Bronze Age until the end of the 2nd millennium AD. During this span of time, Information and Communication Technologies (ICTs) evolved from being mainly recording systems – writing and manuscript production – to being also communication systems, especially after Gutenberg and the invention of printing – to being also processing and producing systems, especially after Turing and the diffusion of computers. Thanks to this evolution, nowadays the most advanced societies highly depend on information-based, intangible assets,

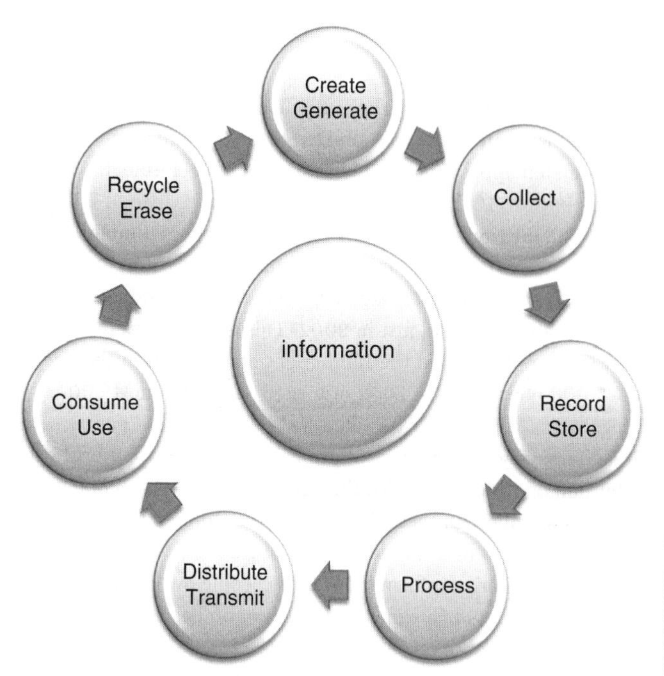


**1. A typical information life cycle**



information-intensive services (especially business and property services, communications, finance and insurance, and entertainment), and information-oriented public sectors (especially education, public administration, and health care). For example, all members of the G7 group – namely Canada, France, Germany, Italy, Japan, the United Kingdom, and the United States of America – qualify as information societies because, in each country, at least 70% of the Gross Domestic Product (GDP) depends on intangible goods, which are information-related, not on material goods, which are the physical output of agricultural or manufacturing processes. Their functioning and growth requires and generates immense amounts of data, more data than humanity has ever seen in its entire history.

## The zettabyte era

In 2003, researchers at Berkeley's School of Information Management and Systems estimated that humanity had accumulated approximately 12 exabytes of data (1 exabyte corresponds to $10^{18}$ bytes or a 50,000-year-long video of DVD quality) in the course of its entire history until the commodification of computers. However, they also calculated that print, film, magnetic, and optical storage media had already produced more than 5 exabytes of data just in 2002. This is equivalent to 37,000 new libraries the size of the Library of Congress. Given the size of the world population in 2002, it turned out that almost 800 megabytes (MB) of recorded data had been produced per person. It is like saying that every newborn baby came into the world with a burden of 30 feet of books, the equivalent of 800 MB of data printed on paper. Of these data, 92% were stored on magnetic media, mostly in hard disks, thus causing an unprecedented 'democratization' of information: more people own more data than ever before. Such exponential escalation has been relentless. According to a more recent study, between 2006 and 2010 the global quantity of digital data will have increased more than six-fold, from 161 exabytes to 988 exabytes. 'Exaflood' is a neologism that has been coined to qualify this tsunami of bytes that is submerging the world. Of course, hundreds of millions of computing machines are constantly employed to keep afloat and navigate through such an exaflood. All the previous numbers will keep growing steadily for the foreseeable future, not least because computers are among the greatest sources of further exabytes. Thanks to them, we are quickly approaching *the age of the zettabyte* (1,000 exabytes). It is a self-reinforcing cycle and it would be unnatural not to feel overwhelmed. It is, or at least should be, a mixed feeling.

ICTs have been changing the world profoundly and irreversibly for more than half a century now, with breathtaking scope and at a neck-breaking pace. On the one hand, they have brought concrete and imminent opportunities of enormous benefit to people's education, welfare, prosperity, and edification, as well as great economic and scientific advantages. Unsurprisingly, the US Department of Commerce and the National Science Foundation have identified Nanotechnology, Biotechnology, Information Technology, and Cognitive Science (NBIC) as research areas of national priority. Note that the three NBC would be virtually impossible without the I. In a comparable move, the EU Heads of States and Governments acknowledged the immense impact of ICTs when they agreed to make the EU 'the most competitive and dynamic knowledge-driven economy by 2010'.

On the other hand, ICTs also carry significant risks and generate dilemmas and profound questions about the nature of reality and of our knowledge of it, the development of information-intensive sciences (e-science), the organization of a fair society (consider the digital divide), our responsibilities and obligations to present and future generations, our understanding of a globalized world, and the scope of our potential interactions with the environment. As a result, they have greatly outpaced our understanding of their conceptual nature and implications, while raising problems whose complexity and global dimensions are rapidly expanding, evolving, and becoming increasingly serious.

A simple analogy may help to make sense of the current situation. The information society is like a tree that has been growing its far-reaching branches much more widely, hastily, and chaotically than its conceptual, ethical, and cultural roots. The lack of balance is obvious and a matter of daily experience in the life of millions of citizens. As a simple illustration, consider identity theft, the use of information to impersonate someone else in order to steal money or get other benefits. According to the Federal Trade Commission, frauds involving identity theft in the US accounted for approximately $52.6 billion of losses in 2002 alone, affecting almost 10 million Americans. The risk is that, like a tree with weak roots, further and healthier growth at the top might be impaired by a fragile foundation at the bottom. As a consequence, today, any advanced information society faces the pressing task of equipping itself with a viable philosophy of information. Applying the previous analogy, while technology keeps growing bottom-up, it is high time we start digging deeper, top-down, in order to expand and reinforce our conceptual understanding of our information age, of its nature, of its less visible implications, and of its impact on human and environmental welfare, and thus give ourselves a chance to anticipate difficulties, identify opportunities, and resolve problems.

The almost sudden burst of a global information society, after a few millennia of relatively quieter gestation, has generated new and disruptive challenges, which were largely unforeseeable only a few decades ago. As the European Group on Ethics in Science and New Technologies (EGE) and the UNESCO Observatory on the Information Society have well documented, ICTs have made the creation, management, and utilization of information, communication, and computational resources vital issues, not only in our understanding of the world and of our interactions with it, but also in our self-assessment and identity. In other words, computer science and ICTs have brought about a *fourth revolution*.



## The fourth revolution

Oversimplifying, science has two fundamental ways of changing our understanding. One may be called *extrovert*, or about the world, and the other *introvert*, or about ourselves. Three scientific revolutions have had great impact both extrovertly and introvertly. In changing our understanding of the external world they also modified our conception of who we are. After Nicolaus Copernicus (1473–1543), the heliocentric cosmology displaced the Earth and hence humanity from the centre of the universe. Charles Darwin (1809–1882) showed that all species of life have evolved over time from common ancestors through natural selection, thus displacing humanity from the centre of the biological kingdom. And following Sigmund Freud (1856–1939), we acknowledge nowadays that the mind is also unconscious and subject to the defence mechanism of repression. So we are not immobile, at the centre of the universe (Copernican revolution), we are not unnaturally separate and diverse from the rest of the animal kingdom (Darwinian revolution), and we are very far from being standalone minds entirely transparent to ourselves, as René Descartes (1596–1650), for example, assumed (Freudian revolution).

One may easily question the value of this classic picture. After all, Freud was the first to interpret these three revolutions as part of a single process of reassessment of human nature and his perspective was blatantly self-serving. But replace Freud with cognitive science or neuroscience, and we can still find the framework useful to explain our intuition that something very significant and profound has recently happened to human self-understanding. Since the 1950s, computer science and ICTs have exercised both an extrovert and an introvert influence, changing not only our interactions with the world but also our self-understanding. In many respects, we are not standalone entities, but rather interconnected informational organisms or *inforgs*, sharing with biological agents and engineered artefacts a global environment ultimately made of information, the infosphere. This is the informational environment constituted by all informational processes, services, and entities, thus including informational agents as well as their properties, interactions, and mutual relations. If we need a representative scientist for the fourth revolution, this should definitely be Alan Turing (1912–1954).

Inforgs should not be confused with the sci-fi vision of a 'cyborged' humanity. Walking around with a Bluetooth wireless headset implanted in our bodies does not seem a smart move, not least because it contradicts the social message it is also meant to be sending: being constantly on call is a form of slavery, and anyone so busy and important should have a personal assistant instead. Being some sort of cyborg is not what people will embrace, but what they will try to avoid. Nor is the idea of inforgs a step towards a genetically modified humanity, in charge of its informational DNA and hence of its future embodiments. This is something that may happen in the future, but it is still too far away, both technically (safely doable) and ethically (morally acceptable), to be seriously discussed at this stage. Rather, the fourth revolution is bringing to light the intrinsically informational nature of human agents. This is more than just saying that individuals have started having a 'data shadow' or digital alter ego, some Mr Hyde represented by their @s, blogs, and https. These obvious truths only encourage us to mistake digital ICTs for merely enhancing technologies. What is in question is a quieter, less sensational, and yet crucial and profound change in our conception of what it means to be an agent and what sort of environment these new agents inhabit. It is a change that is happening not through some fanciful alterations in our bodies, or some science-fictional speculations about our posthuman condition but, far more seriously and realistically, through a radical transformation of our understanding of reality and of ourselves. A good way to explain it is by relying on the distinction between *enhancing* and *augmenting* appliances.

Enhancing appliances, like pacemakers, spectacles, or artificial limbs, are supposed to have interfaces that enable the appliance to be attached to the user's body ergonomically. It is the beginning of the cyborg idea. Augmenting appliances have instead interfaces that allow communication between different possible worlds. For example: on one side, there is the human user's everyday habitat, the outer world, or reality, as it affects the agent inhabiting it; and on the other side, there are the dynamic, watery, soapy, hot, and dark world of the dishwasher; the equally watery, soapy, hot, and dark but also spinning world of the washing machine; or the still, aseptic, soapless, cold, and potentially luminous world of the refrigerator. These robots can be successful because they have their environments 'wrapped' and tailored around their capacities, not vice versa. This is why it would be a silly idea to try to build a droid, like *Star Wars*' C3PO, in order to wash dishes in the sink exactly in the same way as a human agent would. Now, ICTs are not enhancing or augmenting in the sense just explained. They are radically transforming devices because they engineer environments that the user is then enabled to enter through (possibly friendly) gateways, experiencing a form of initiation. There is no term for this radical form of re-engineering, so we may use *re-ontologizing* as a neologism to refer to a very radical form of re-engineering, one that not only designs, constructs, or structures a system (e.g. a company, a machine, or some artefact) anew, but that fundamentally transforms its intrinsic nature, that is, its ontology. In this sense, ICTs are not merely re-engineering but actually re-ontologizing our world. Looking at the history of the mouse (http://sloan.stanford.edu/mousesite/), for example, one discovers that our technology has not only adapted to, but also educated, us as users. Douglas Engelbart (born 1925) once told me that, when he was refining his most famous invention, the mouse, he even experimented with placing it under the desk, to be operated with one's leg, in order to leave the user's hands free. Human–Computer Interaction is a symmetric relation.

To return to our distinction, while a dishwasher interface is a panel through which the machine enters into the user's world, a digital interface is a gate through which a user can be present in cyberspace. This simple but fundamental difference underlies the many spatial metaphors of 'virtual reality', 'being online', 'surfing the web', 'gateway', and so forth. It follows that we are witnessing an epochal, unprecedented migration of humanity from its ordinary habitat to the infosphere itself, not least because the latter is absorbing the former. As a result, humans will be inforgs among other (possibly artificial) inforgs and agents operating in an environment that is friendlier to informational creatures. Once digital immigrants like us are replaced by digital natives like our children, the e-migration will become complete and future generations will increasingly feel deprived, excluded, handicapped, or poor whenever they are disconnected from the infosphere, like fish out of water.

What we are currently experiencing is therefore a *fourth revolution*, in the process of dislocation and reassessment of our fundamental nature and role in the universe. We are modifying our everyday perspective on the ultimate nature of reality, that is, our metaphysics, from a materialist one, in which physical objects and processes play a key role, to an informational one. This shift means that objects and processes are de-physicalized in the sense that they tend to be seen as support-independent (consider a music file). They are typified, in the sense that an instance of an object (my copy of a music file) is as good as its type (your music file of which my copy is an instance). And they are assumed to be by default perfectly clonable, in the sense that my copy and your original become interchangeable. Less stress on the physical nature of objects and processes means that the right of usage is perceived to be at least as important as the right to ownership. Finally, the criterion for existence – what it means for something to exist – is no longer being actually immutable (the Greeks thought that only that which does not change can be said to exist fully), or being potentially subject to perception (modern philosophy insisted on something being perceivable empirically through the five senses in order to qualify as existing), but being potentially subject to interaction, even if intangible. To be is to be interactable, even if the interaction is only indirect. Consider the following examples.

In recent years, many countries have followed the US in counting acquisition of software not as a current business expense but as an investment, to be treated as any other capital input that is repeatedly used in production over time. Spending on software now regularly contributes to GDPs. So software is acknowledged to be a (digital) good, even if somewhat intangible. It should not be too difficult to accept that virtual assets too may represent important investments. Or take the phenomenon of so-called 'virtual sweatshops' in China. In claustrophobic and overcrowded rooms, workers play online games, like *World of Warcraft* or *Lineage*, for up to 12 hours a day, to create virtual goods, such as characters, equipments, or in-game currency, which can then be sold to other players. At the time of writing, End User License Agreements (EULA, this is the contract that every user of commercial software accepts by installing it) of massively multiplayer online role-playing games (MMORPG) such as *World of Warcraft* still do not allow the sale of virtual assets. This would be like the EULA of MS-Word withholding from users the ownership of the digital documents created by means of the software. The situation will probably change, as more people invest hundreds and then thousands of hours building their avatars and assets. Future generations will inherit digital entities that they will want to own. Indeed, although it was forbidden, there used to be thousands of virtual assets on sale on eBay. Sony, more aggressively, offers a 'Station Exchange', an official auction service that 'provides players a secure method of buying and selling [in dollars, my specification] the right to use in game coin, items and characters in accordance with SOE's licence agreement, rules, and guidelines' (http://stationexchange.station.sony.com/). Once ownership of virtual assets has been legally established, the next step is to check for the emergence of property litigations. This is already happening: in May 2006, a Pennsylvania lawyer sued the publisher of *Second Life* for allegedly having unfairly confiscated tens of thousands of dollars' worth of his virtual land and other property. Insurances that provide protection against risks to avatars may follow, comparable to the pet insurances one can buy at the local supermarket. Again, *World of Warcraft* provides an excellent example. With almost 12 million monthly subscribers (2009), it is currently the world's largest MMORPG and would rank 71st in the list of 221 countries and dependent territories ordered according to population. Its users, who (will) have spent billions of man-hours constructing, enriching, and refining their digital properties, will be more than willing to spend a few dollars to insure them.

ICTs are actually creating a new informational environment in which future generations will live most of their time. On average, Britons, for example, already spend more time online than watching TV, while American adults already spend the equivalent of nearly five months a year inside the infosphere. Such population is quickly ageing. According to the Entertainment Software Association, for example, in 2008 the average game player was 35 years old and had been playing games for 13 years, the average age of the most frequent game purchaser is 40 years old, and 26% of Americans over the age of 50 played video games, an increase from 9% in 1999.



## Life in the infosphere

Despite some important exceptions (e.g. vases and metal tools in ancient civilizations, engravings and then books after Gutenberg), it was the Industrial Revolution that really marked the passage from a world of unique objects to a world of types of objects, all perfectly reproducible as identical to each other, therefore indiscernible, and hence dispensable because replaceable without any loss in the scope of interactions that they allow. When our ancestors bought a horse, they bought *this* horse or *that* horse, not 'the' horse. Today, we find it obvious that two automobiles may be virtually identical and that we are invited to buy a model rather than an individual 'incarnation' of it. Indeed, we are fast moving towards a commodification of objects that considers repair as synonymous with replacement, even when it comes to entire buildings. This has led, by way of compensation, to a prioritization of informational *branding* and of *re-appropriation*: the person who puts a sticker on the window of her car, which is otherwise perfectly identical to thousands of others, is fighting a battle in support of her individualism. The information revolution has further exacerbated this process. Once our window-shopping becomes Windows-shopping, and no longer means walking down the street but browsing through the Web, our sense of personal identity starts being eroded as well. Instead of individuals as unique and irreplaceable entities, we become mass-produced, anonymous entities among other anonymous entities, exposed to billions of other similar informational organisms online. So we self-brand and re-appropriate ourselves in the infosphere by using blogs and Facebook entries, homepages, YouTube videos, and flickr albums. It is perfectly reasonable that *Second Life* should be a paradise for fashion enthusiasts of all kinds: not only does it provide a new and flexible platform for designers and creative artists, it is also the right context in which users (avatars) intensely feel the pressure to obtain visible signs of self-identity and personal tastes. Likewise, there is no inconsistency between a society so concerned about privacy rights and the success of services such as Facebook. We use and expose information about ourselves to become less informationally anonymous. We wish to maintain a high level of informational privacy, almost as if that were the only way of saving a precious capital that can then be publicly invested by us in order to construct ourselves as individuals discernible by others.

Processes such as the ones I have just sketched are part of a far deeper metaphysical drift caused by the information revolution. During the last decade or so, we have become accustomed to conceptualizing our life online as a mixture between an evolutionary adaptation of human agents to a digital environment, and a form of postmodern, neo-colonization of that space by us. Yet the truth is that ICTs are as much changing our world as they are creating new realities. The threshold between *here* (*analogue*, *carbon-based*, *off-line*) and *there* (*digital*, *silicon-based*, *online*) is fast becoming blurred, but this is as much to the advantage of the latter as it is of the former. The digital is spilling over into the analogue and merging with it. This recent phenomenon is variously known as 'Ubiquitous Computing', 'Ambient Intelligence', 'The Internet of Things', or 'Web-augmented things'.

The increasing informatization of artefacts and of whole (social) environments and life activities suggests that soon it will be difficult to understand what life was like in pre-informational times (to someone who was born in 2000, the world will always have been wireless, for example) and, in the near future, the very distinction between online and offline will disappear. The common experience of driving a car while following the instructions of a Global Positioning System clarifies how pointless asking whether one is online has become. To put it dramatically, the infosphere is progressively absorbing any other space. In the (fast-approaching) future, more and more objects will be *ITentities* able to learn, advise, and communicate with each other. A good example (but it is only an example) is provided by Radio Frequency IDentification (RFID) tags, which can store and remotely retrieve data from an object and give it a unique identity, like a barcode. Tags can measure 0.4 millimetres square and are thinner than paper. Incorporate this tiny microchip in everything, including humans and animals, and you have created *ITentities*. This is not science fiction. According to a report by market research company InStat, the worldwide production of RFID will have increased more than 25-fold between 2005 and 2010 and reached 33 billion. Imagine networking these 33 billion ITentities together with all the hundreds of millions of PCs, DVDs, iPods, and other ICT devices available and you see that the infosphere is no longer 'there' but 'here' and it is here to stay. Your Nike Sensor and iPod already talk to each other (http://www.apple.com/ipod/nike/).

At present, older generations still consider the space of information as something one logs-in to and logs-out from. Our view of the world (our metaphysics) is still modern or Newtonian: it is made of 'dead' cars, buildings, furniture, clothes, which are non-interactive, irresponsive, and incapable of communicating, learning, or memorizing. But in advanced information societies, what we still experience as the world offline is bound to become a fully interactive and more responsive environment of wireless, pervasive, distributed, *a2a* (anything to anything) information processes, that works *a4a* (anywhere for anytime), in real time. Such a world will first gently invite us to understand it as something 'a-live' (artificially live). This *animation* of the world will then, paradoxically, make our outlook closer to that of pre-technological cultures, which interpreted all aspects of nature as inhabited by teleological forces.



This leads to a reconceptualization of our metaphysics in informational terms. It will become normal to consider the world as part of the infosphere, not so much in the dystopian sense expressed by a *Matrix*-like scenario, where the 'real reality' is still as hard as the metal of the machines that inhabit it; but in the evolutionary, hybrid sense represented by an environment such as New Port City, the fictional, post-cybernetic metropolis of *Ghost in the Shell*. The infosphere will not be a virtual environment supported by a genuinely 'material' world behind; rather, it will be the world itself that will be increasingly interpreted and understood informationally, as part of the infosphere. At the end of this shift, the infosphere will have moved from being a way to refer to the space of information to being synonymous with reality. This is the sort of informational metaphysics that we may find increasingly easy to embrace.

As a consequence of such transformations in our ordinary environment, we shall be living in an infosphere that will become increasingly *synchronized* (time), *delocalized* (space), and *correlated* (interactions). Previous revolutions (especially the agricultural and the industrial ones) created macroscopic transformation in our social structures and architectural environments, often without much foresight. The information revolution is no less dramatic. We shall be in trouble if we do not take seriously the fact that we are constructing the new environment that will be inhabited by future generations. At the end of this volume, we shall see that we should probably be working on an ecology of the infosphere, if we wish to avoid foreseeable problems. Unfortunately, it will take some time and a whole new kind of education and sensitivity to realize that the infosphere is a common space, which needs to be preserved to the advantage of all. One thing seems indubitable though: the digital divide will become a chasm, generating new forms of discrimination between those who can be denizens of the infosphere and those who cannot, between insiders and outsiders, between information-rich and information-poor. It will redesign the map of worldwide society, generating or widening generational, geographic, socio-economic, and cultural divides. But the gap will not be reducible to the distance between industrialized and developing countries, since it will cut across societies. We are preparing the ground for tomorrow's digital slums.

# Chapter 2
# The language of information

Information is a conceptual labyrinth, and in this chapter we shall look at its general map, with the purpose of finding our bearings. Figure 2 summarizes the main distinctions that are going to be introduced. Some areas will be explored in more depth in the following chapters.

Navigating through the various points in the map will not make for a linear journey, so using a few basic examples to illustrate the less obvious steps will help to keep our orientation. Here is one to which we shall return often.

It is Monday morning. John turns on the ignition key of his car, but nothing happens: the engine does not even cough. The silence of the engine worries him. Looking more carefully, he notices that the low-battery indicator is flashing. After a few more unsuccessful attempts, he gives up and calls the garage. Over the phone, he explains that, last night, his wife forgot to switch off the car's lights – it is a lie, John did, but he is too ashamed to admit it – and now the battery is flat. The mechanic tells John that he should look at the car's operation manual, which explains how to use jump leads to start the engine. Luckily, John's neighbour has everything he needs. He reads the manual, looks at the illustrations, speaks to his

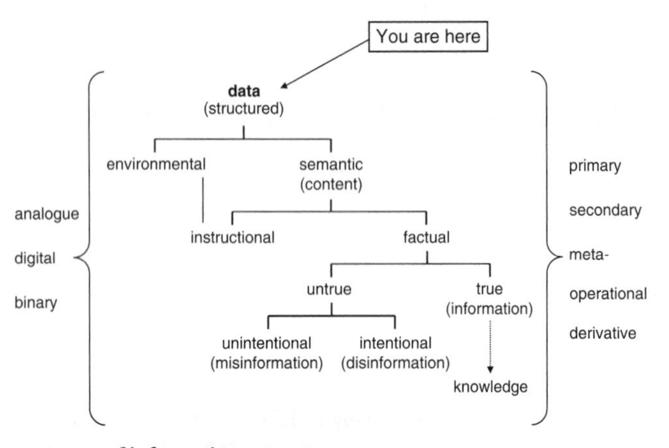


**2. A map of information concepts**



neighbour, follows the instructions, solves the problem, and finally drives to the office.

This everyday episode will be our 'fruit fly', as it provides enough details to illustrate the many ways in which we understand information. Our first step will now be to define information in terms of data.

## The data-based definition of information

Over the past decades, it has become common to adopt a *General Definition of Information* (GDI) in terms of *data* + *meaning*. GDI has become an operational standard, especially in fields that treat data and information as reified entities, that is, stuff that can be manipulated (consider, for example, the now common expressions 'data mining' and 'information management'). A straightforward way of formulating GDI is as a tripartite definition (Table 1):

According to (GDI.1), information is made of data. In (GDI.2), 'well formed' means that the data are rightly put together,

**Table 1. The General Definition of Information (GDI)**

GDI) $\sigma$ is an instance of information, understood as semantic content, if and only if:
GDI.1) $\sigma$ consists of $n$ *data*, for $n \geq 1$;
GDI.2) the data are *well formed*;
GDI.3) the well-formed data are *meaningful*.

according to the rules (*syntax*) that govern the chosen system, code, or language being used. Syntax here must be understood broadly, not just linguistically, as what determines the form, construction, composition, or structuring of something. Engineers, film directors, painters, chess players, and gardeners speak of syntax in this broad sense. In our example, the car's operation manual may show a two-dimensional picture of how to jump-start a car. This pictorial syntax (including the linear perspective that represents space by converging parallel lines) makes the illustration potentially meaningful to the user. Still relying on the same example, the actual battery needs to be connected to the engine in a correct way to function properly: this is still syntax, in terms of correct physical architecture of the system (thus a disconnected battery is a syntactic problem). And of course, the conversation John carries on with his neighbour follows the grammatical rules of English: this is syntax in the ordinary linguistic sense.

Regarding (GDI.3), this is where semantics finally occurs. 'Meaningful' means that the data must comply with the meanings (*semantics*) of the chosen system, code, or language in question. Once again, semantic information is not necessarily linguistic. For example, in the case of the car's operation manual, the illustrations are supposed to be visually meaningful to the reader.

How data can come to have an assigned meaning and function in a semiotic system like a natural language is one of the hardest questions in semantics, known as the *symbol grounding problem*. Luckily, it can be disregarded here. The only point worth clarifying is that data constituting information can be meaningful independently of an informee. Consider the following example. The Rosetta Stone contains three translations of a single passage, in Egyptian hieroglyphic, Egyptian Demotic, and classical Greek languages. Before its discovery, Egyptian hieroglyphics were already regarded as information, even if their meaning was beyond the comprehension of any interpreter. The discovery of an interface between Greek and Egyptian did not affect the semantics of the hieroglyphics but only its *accessibility*. This is the reasonable sense in which one may speak of meaningful data being embedded in information-carriers independently of any informee. It is very different from the stronger thesis, according to which data could also have their own semantics independently of an intelligent *producer/informer*. This is also known as *environmental information*, but, before discussing it, we need to understand much better the nature of data.



## Understanding data

A good way to uncover the most fundamental nature of data is by trying to understand what it means to erase, damage, or lose them. Imagine the page of a book written in a language unknown to us. Suppose the data are in the form of pictograms. The regular patterns suggest the compliance with some structural syntax. We have all the data, but we do not know their meaning, hence we have no information yet. Let us now erase half of the pictograms. One may say that we have halved the data as well. If we continue in this process, when we are left with only one pictogram we might be tempted to say that data require, or may be identical with, some sort of representations. But now let us erase that last pictogram too. We are left with a white page, and yet not entirely without data. For the presence of a white page is still a datum, as long as there is a difference between the white page and the page on which something is or could be written. Compare this to the common phenomenon of 'silent assent': silence, or the lack of perceivable data, can be as much a datum as the presence of some noise, exactly like the zeros of a binary system. Recall in our example John's concern when he did not hear any sound coming from his car's engine. That lack of noise was informative. The fact is that a genuine, complete erasure of all data can be achieved only by the elimination of all possible differences. This clarifies why a datum is ultimately reducible to a *lack of uniformity*. Donald MacCrimmon MacKay (1922–1987) highlighted this important point when he wrote that 'information is a distinction that makes a difference'. He was followed by Gregory Bateson (1904–1980), whose slogan is better known, although less accurate: 'In fact, what we mean by information – the elementary unit of information – is a difference which makes a difference'. More formally, according to the *diaphoric interpretation* (*diaphora* is the Greek word for 'difference'), the general definition of a datum is:

> Dd) datum $=_{\text{def.}}$ $x$ being distinct from $y$, where $x$ and $y$ are two uninterpreted variables and the relation of 'being distinct', as well as the domain, are left open to further interpretation.

This definition of data can be applied in three main ways.

First, data can be lacks of uniformity in the real world. There is no specific name for such 'data in the wild'. One may refer to them as *dedomena*, that is, 'data' in Greek (note that our word 'data' comes from the Latin translation of a work by Euclid entitled *Dedomena*). Dedomena are not to be confused with *environmental information*, which will be discussed later in this chapter. They are pure data, that is, data before they are interpreted or subject to cognitive processing. They are not experienced directly, but their presence is empirically inferred from, and required by, experience, since they are what has to be there in the world for our information to be possible at all. So dedomena are whatever lack of uniformity in the world is the source of (what looks to informational organisms like us as) data, e.g. a red light against a dark background. I shall return to this point in Chapter 5, where we shall see that some researchers have been able to accept the thesis that there can be no information without data while rejecting the thesis that information must have a material nature.

Second, data can be lacks of uniformity between (the perception of) at least two physical states of a system or *signals*. Examples include a higher or lower charge in a battery, a variable electrical signal in a telephone conversation, or the dot and the line in the Morse alphabet.

Finally, data can be lacks of uniformity between two *symbols*, for example the letters B and P in the Latin alphabet.

Depending on one's interpretation, the *dedomena* in (1) may be either identical with, or what makes possible *signals* in (2), and signals in (2) are what make possible the coding of *symbols* in (3).



The dependence of information on the occurrence of syntactically well-formed data, and of data on the occurrence of differences variously implementable physically, explain why information can

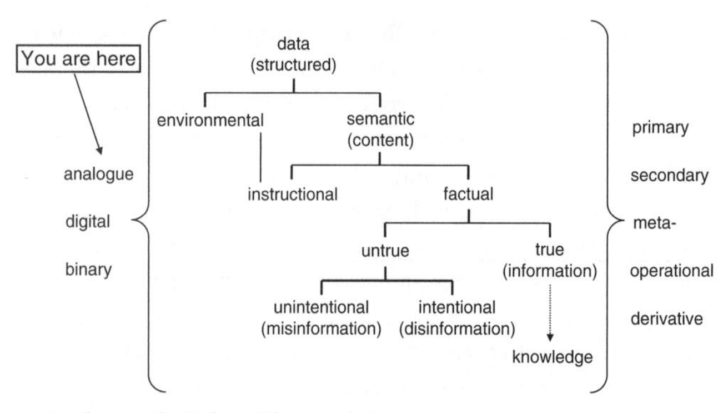


**3. Analogue, digital, and binary data**

so easily be decoupled from its support. The actual *format*, *medium*, and *language* in which data, and hence information, are encoded is often irrelevant and disregardable. In particular, the same data/information may be printed on paper or viewed on a screen, codified in English or in some other language, expressed in symbols or pictures, be analogue or digital. The last distinction is the most important and deserves some clarification.

## Analogue versus digital data

Analogue data and the systems that encode, store, process, or transmit them vary continuously. For example, vinyl records are analogue because they store mechanical, continuous data that correspond to the recorded sounds. On the contrary, digital data and the related systems vary discretely between different states, e.g. on/off or high/low voltage. For example, compact discs are digital because they store sounds by transforming them as series of pits (indentations) and lands (the areas between pits). They *encode* rather than just *record* information.



Our understanding of the universe is firmly based not only on digital, discrete, or grainy ideas – the natural numbers, the heads or tails of a coin, the days of the week, the goals scored by a football team, and so forth – but also on many analogue, continuous, or smooth ideas – the intensity of a pain or pleasure, the real numbers, continuous functions, differential equations, waves, force fields, the continuum of time. Computers are usually seen as digital or discrete information systems, but this is not entirely correct, for two reasons. As Turing himself remarked,

> The digital computers [. . .] may be classified amongst the 'discrete state machines', these are the machines which move by sudden jumps or clicks from one quite definite state to another. These states are sufficiently different for the possibility of confusion between them to be ignored. Strictly speaking there are no such machines.

Everything really moves continuously. But there are many kinds of machine, which can profitably be thought of as being discrete state machines.

And there are analogue computers. These perform calculations through the interaction of continuously varying physical phenomena, such as the shadow cast by the gnomon on the dial of a sundial, the approximately regular flow of sand in an hourglass or of water in a water clock, and the mathematically constant swing of a pendulum. Clearly, it is not the use of a specific substance or reliance on a specific physical phenomenon that makes an information system analogue, but the fact that its operations are directly determined by the measurement of continuous, physical transformations of whatever solid, liquid, or gaseous matter is employed. There are analogue computers that use continuously varying voltages and a Turing machine (the logically idealized model of our personal computers) is a digital computer but may not be electrical. Given their physical nature, analogue computers operate in real time (i.e. time corresponding to time in the real world) and therefore can be used to monitor and control events as they happen, in a 1:1 relation between the time of the event and the time of computation (think of the hourglass). However, because of their nature, analogue computers cannot be general-purpose machines but can only perform as necessarily specialized devices. The advantage is that analogue data are highly resilient: a vinyl record can be played again and again, even if it is scratched.

## Binary data

Digital data are also called binary data because they are usually encoded by means of combinations of only two symbols called *bits* (*b*inary dig*its*), as strings of 0s and 1s comparable to the dots and dashes in the Morse code. For example, in binary notation the number three is written 11 (see Table 2). Since the value of any position in a binary number increases by the power of 2 (doubles) with each move from right to left (i.e. . . . 16, 8, 4, 2, 1; note that it could have been 1, 2, 4, 8, 16, and so on, but the binary system pays due homage to the Arabic language and moves from right to left) 11 means $(1 \times 2) + (1 \times 1)$, which adds up to three in the decimal system. Likewise, if one calculates the binary version of 6, equivalent to $(1 \times 4) + (1 \times 2) + (0 \times 1)$ one can see that it can only be 110.

**Table 2. Decimal and binary notations of positive integers**

| | | | Decimal Notation | | |
|---|---|---|---|---|---|
| | ... | $10^3 = 1000$ | $10^2 = 100$ | $10^1 = 10$ | $10^0 = 1$ |
| one apple | | | | | 1 |
| two apples | | | | | 2 |
| ... | | | | | |
| six apples | | | | | 6 |
| ... | | | | | |
| thirteen apples | | | | 1 | 3 |
| ... | | | | | |
| | | | **Binary Notation** | | |
| | ... | $2^3 = 8$ | $2^2 = 4$ | $2^1 = 2$ | $2^0 = 1$ |
| one apple | | | | | 1 |
| two apples | | | | 1 | 0 |
| ... | | | | | |
| six apples | | | 1 | 1 | 0 |
| ... | | | | | |
| thirteen apples | | 1 | 1 | 0 | 1 |
| ... | | | | | |

A *bit* is the smallest unit of information, nothing more than the presence or absence of a signal, a 0 or a 1. A series of 8 bits forms a *byte* (*by* eigh*t*), and by combining bytes it becomes possible to generate a table of 256 ($2^8$) characters. Each character of data can then be stored as a pattern of 8 bits. The most widely used binary code is known as ASCII (American Standard Code for Information Interchange), which relies on only 7 bits out of 8 and therefore consists of a table of 128 ($2^7$) characters. Here is how a computer spells 'GOD' in binary: 010001110100111101000100 (Table 3):

**Table 3. Example of binary encoding**

| | | | | | | | | |
|---|---|---|---|---|---|---|---|---|
| G | off = 0 | on = 1 | off = 0 | off = 0 | off = 0 | on = 1 | on = 1 | on = 1 |
| O | off = 0 | on = 1 | off = 0 | off = 0 | on = 1 | on = 1 | on = 1 | on = 1 |
| D | off = 0 | on = 1 | off = 0 | off = 0 | off = 0 | on = 1 | off = 0 | off = 0 |



Quantities of bytes are then calculated according to the binary system:

- 1 Kilobyte (KB) = $2^{10}$ = 1,024 bytes
- 1 Megabyte (MB) = $2^{20}$ = 1,048,576 bytes
- 1 Gigabyte (GB) = $2^{30}$ = 1,073,741,824 bytes
- 1 Terabyte (TB) = $2^{40}$ = 1,099,511,627,776 bytes

and so forth.

This is why the precise size of the random-access memory (RAM) of a computer, for example, is never a round number.

The binary system of data encoding has at least three advantages. First, bits can equally well be represented semantically (meaning True/False), logico-mathematically (standing for 1/0), and physically (transistor = On/Off; switch = Open/Closed; electric circuit = High/Low voltage; disc or tape = Magnetized/

Unmagnetized; CD = presence/absence of pits, etc.), and hence provide the common ground where semantics, mathematical logic, and the physics and engineering of circuits and information theory can converge.

This means (second advantage) that it is possible to construct machines that can recognize bits physically, behave logically on the basis of such recognition, and therefore manipulate data in ways which we find meaningful. This is a crucial fact. The only glimpse of intelligence everyone is ready to attribute to a computer uncontroversially concerns the capacity of its devices and circuits to discriminate between binary data. If a computer can be said to perceive anything at all, this is the difference between a high and a low voltage according to which its circuits are then programmed to behave. The odd thing is that this may be somewhat true of biological systems as well, as we shall see in Chapter 6.

Finally, since digital data normally have only two states, such *discrete variation* means that a computer will hardly ever get confused about what needs to be processed, unlike an analogue machine, which can often perform unsatisfactorily or imprecisely. Above all, a digital machine can recognize if some data are incomplete and hence recover, through mathematical calculations, data that may have got lost if there is something literally odd about the quantity of bits it is handling.

## Types of data/information

Information can consist of different types of data. Five classifications are quite common, although the terminology is not yet standard or fixed. They are not mutually exclusive, and one should not understand them as rigid: depending on circumstances, on the sort of analysis conducted, and on the perspective adopted, the same data may fit different classifications.

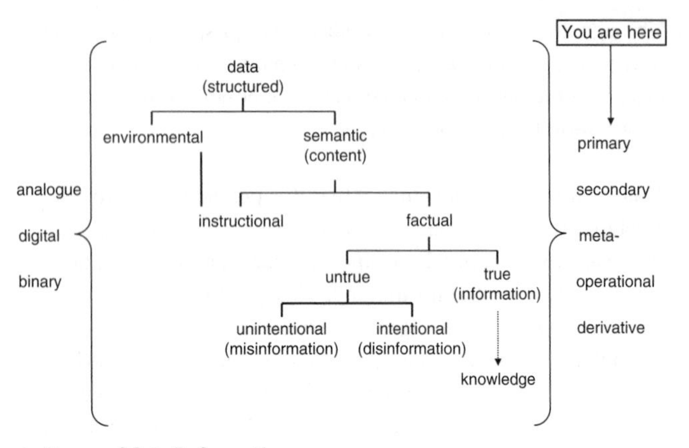


**4. Types of data/information**



## Primary data

These are the principal data stored in a database, for example a simple array of numbers in a spreadsheet, or a string of zeroes and ones. They are the data an information-management system – such as the one indicating that the car's battery needs to be recharged – is generally designed to convey to the user in the first place, in the form of information. Normally, when speaking of data, and of the corresponding information they constitute, one implicitly assumes that *primary* data/information is what is in question. So, by default, the red light of the low-battery indicator flashing is assumed to be an instance of primary data conveying primary information, not some secrete message for a spy.

## Secondary data

These are the converse of primary data, constituted by their absence. Recall how John first suspected that the battery was flat: the engine failed to make any noise, thus providing secondary information about the flat battery. Likewise, in *Silver Blaze*, Sherlock Holmes solves the case by noting something that has escaped everybody else: the unusual silence of the dog. Clearly, silence may be very informative. This is a peculiarity of information: its absence may also be informative. When it is, the point is stressed by speaking of *secondary information*.

### Metadata

These are indications about the nature of some other (usually primary) data. They describe properties such as location, format, updating, availability, usage restrictions, and so forth. Correspondingly, *metainformation* is information about the nature of information. The copyright note on the car's operation manual is a simple example.

### Operational data

These are data regarding the operations of the whole data system and the system's performance. Correspondingly, *operational information* is information about the dynamics of an information system. Suppose the car has a yellow light that, when flashing, indicates that the car checking system is malfunctioning. The fact that the yellow light is on may indicate that the low-battery indicator (the red light flashing) is not working properly, thus undermining the hypothesis that the battery is flat.

### Derivative data

These are data that can be extracted from some data whenever the latter are used as indirect sources in search of patterns, clues, or inferential evidence about other things than those directly addressed by the data themselves, for example for comparative and quantitative analyses. As it is difficult to define this category precisely, let me rely on our familiar example. Credit cards notoriously leave a trail of derivative information. From John's credit card bill, concerning his purchase of petrol in a specific petrol station, one may obtain the derivative information of his whereabouts at a given time.

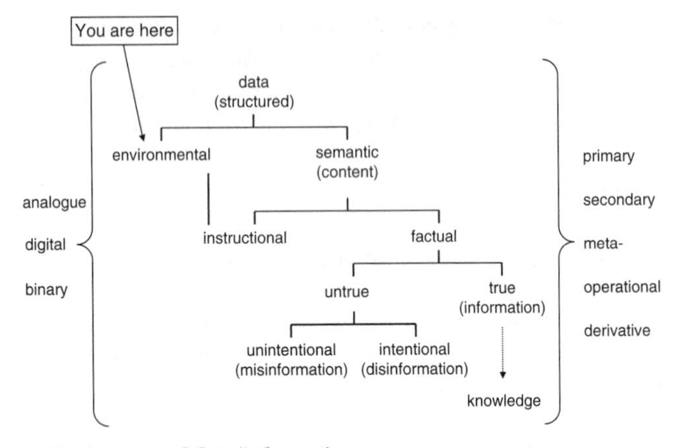


**5. Environmental data/information**



We are now ready to look at environmental information.

## Environmental information

We speak of *environmental information* when we wish to stress the possibility that data might be meaningful independently of an intelligent *producer/informer*. One of the most often cited examples of environmental information is the series of concentric rings visible in the wood of a cut tree trunk, which may be used to estimate its age. Viewers of *CSI: Crime Scene Investigation*, the crime television series, will also be well acquainted with bullet trajectories, blood spray patterns, organ damages, fingerprints, and other similar evidence. Yet 'environmental' information does not need to be *natural*. Going back to our example, when John turned the ignition key, the red light of the low-battery indicator flashed. This *engineered* signal too can be interpreted as an instance of environmental information. The latter is normally defined relative to an observer (an informational organism or informee), who relies on it instead of having direct access to the original data in themselves. It follows that environmental information requires two systems, let us call them $a$ and $b$, which are linked in such a way that the fact that $a$ has a particular feature $F$ is correlated to the fact that $b$ has a particular feature $G$, so that this connection between the two features tells the observer that $b$ is $G$. In short:

**Table 4. Environmental information**

| |
|---|
| Environmental information $=_{def.}$ two systems $a$ and $b$ coupled in such a way that $a$'s being (of type, or in state) $F$ is correlated to $b$ being (of type, or in state) $G$, thus carrying for the observer of $a$ the information that $b$ is $G$. |

The correlation in Table 4 follows some law or rule. A *natural* example is provided by litmus. This is a biological colouring matter from lichens that is used as an acid/alkali indicator because it turns red in acid solutions and blue in alkaline solutions. Following the definition of environmental information, we can see that litmus ($a$) and the tested solution ($b$) are coupled in such a way that litmus turning red ($a$ being in state $F$) is correlated to the solution being acid ($b$ being of type $G$), thus carrying the information for the observer of the litmus ($a$) that the solution is acid ($b$ is $G$). Our car example provides an *engineered* case: the low-battery indicator ($a$) flashing ($F$) is triggered by, and hence is informative about, the battery ($b$) being flat ($G$).

We may be so used to seeing the low-battery indicator flashing as carrying the information that the battery is flat that we may find it hard to distinguish, with sufficient clarity, between environmental and semantic information: the red light flashing *means* that the battery is low. However, it is important to stress that environmental information may require or involve no semantics at all. It may consist of networks or patterns of correlated data understood as mere physical differences. Plants, animals, and mechanisms – e.g., a sunflower, an amoeba, or a photocell – are certainly capable of making practical use of environmental information even in the absence of any semantic processing of *meaningful* data (see Chapter 6).

## Information as semantic content

When data are well formed and meaningful, the result is also known as *semantic content*. Information, understood as semantic content, comes in two main varieties: *instructional* and *factual*. In our example, one may translate the red light flashing into semantic content in two senses:

(a) as a piece of instructional information, conveying the need for a specific action, e.g. recharging or replacing of the flat battery; and
(b) as a piece of factual information, representing the fact that the battery is flat.



Chapter 4 will be primarily about (b), so this chapter ends with a discussion of (a).

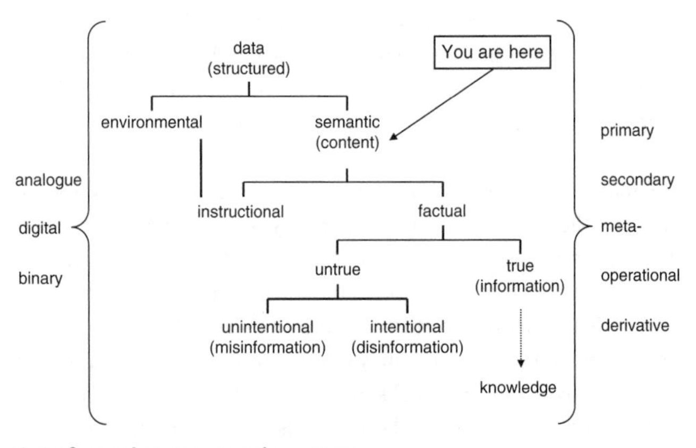


**6. Information as semantic content**

Instructional information can be a type of environmental information or of semantic content, depending on whether meaning is a required feature. For example, the logic gates in the motherboard of a computer merely channel the electric voltage, which we may then interpret in terms of instructional information (logic instructions), such as 'if... then'. In this case, there is no semantics involved at the level of the gates. The car's operation manual, on the contrary, provides *semantic* instructional information, either imperatively – in the form of a recipe: first do this, then do that – or conditionally – in the form of some inferential procedure: if such and such is the case do this, otherwise do that.

Whether environmental or semantic, instructional information is not *about* a situation, a fact, or a state of affairs $w$ and does not model, or describe, or represent $w$. Rather, it is meant to (contribute to) bring about $w$. Compare the difference between 'the water in the kettle has just boiled', which is an instance of factual semantic information, and the process caused by the steam when it heats up the bimetallic strip enough to break the circuit of electricity flowing through the element inside the kettle, which might be interpreted in terms of instructional information. In our example, when the mechanic tells John, over the phone, to connect a charged battery to the flat battery of his car, the information John receives is not factual, but instructional. We shall return to environmental instructional information in Chapter 6, when discussing biological information. Here, let us concentrate on the semantic aspects.

There are many plausible contexts in which a stipulation ('let the value of × be 3' or 'suppose we genetically engineer a unicorn'), an invitation ('you are cordially invited to the college party'), an order ('close the window!'), an instruction ('to open the box turn the key'), a game move ('1.e2-e4 c7-c5' at the beginning of a chess game) may be correctly qualified as kinds of semantic instructional information. The printed score of a musical composition or the digital files of a program may also be counted as typical cases of instructional information. Such semantic instances of instructional information have to be at least potentially meaningful (interpretable) to count as information. Finally, there are performative contexts in which we do things with words, such as christening (e.g. 'this ship is now called *HMS The Informer*') or programming (e.g. as when deciding the type of a variable). In these cases, factual (descriptive) information acquires an instructional value.

As readers of Harry Potter might suspect, the two types of semantic information (instructional and factual) may come together in magic spells, where semantic representations of $x$ may be supposed to provide some instructional power and control over $x$, wrongly in real life, rightly in Harry Potter's adventures. Nevertheless, as a test, one should remember that instructional information cannot be correctly qualified as true or false. In the example, it would be silly to ask whether the information 'only use batteries with the same rated voltage' is true. Likewise, stipulations, invitations, orders, instructions, game moves, and software cannot be true or false.

Semantic information is often supposed to be *declarative* or *factual*. Factual information such as a train timetable, a bank account statement, a medical report, a note saying that tomorrow the library will not be open, and so forth, may be sensibly qualified as true or false. *Factual semantic content* is therefore the most common way in which information is understood and also one of the most important, since information as true semantic content is a necessary condition for knowledge. Because of this key role, Chapter 4 is entirely dedicated to it. Before dealing with it, however, we need to complete our exploration of the concepts of information that require neither meaning nor truth. This is the task of the next chapter, dedicated to the mathematical theory of communication, also known as information theory.

# Chapter 3
# Mathematical information

Some features of information are intuitively quantifiable. A broadband network can carry only a maximum amount of information per second. A computer has a hard disk which can contain only a finite amount of information. More generally, we are used to information being *encoded*, *transmitted*, and *stored* in specific quantities, like physical signals. We also expect it to be *additive* like biscuits and coins: if I give you information $a +$ information $b$, I have given you information $a + b$. And we understand information as being never *negative*: like probabilities and interest rates information cannot go below zero, unlike my bank account or the temperature in Oxford. Consider our example. When John asks a question to his neighbour, the worst scenario is that he will receive no answer or a wrong answer, which would leave him with zero new information.

These and other quantitative properties of information are investigated by many successful mathematical approaches. The *mathematical theory of communication* (MTC) is by far the most important, influential, and widely known. The name for this branch of probability theory comes from Claude Shannon's seminal work. Shannon pioneered the field of mathematical studies of information and obtained many of its principal results, even though he acknowledged the importance of previous work done by other researchers and colleagues at Bell laboratories. After Shannon, MTC became known as *information theory*. Today, Shannon is considered 'the father of information theory', and the kind of information MTC deals with is often qualified as Shannon information. The term 'information theory' is an appealing but unfortunate label, which continues to cause endless misunderstandings. Shannon came to regret its widespread popularity, and I shall avoid it in this context.

MTC is the theory that lies behind any phenomenon involving data encoding and transmission. As such, it has had a profound impact on the analyses of the various kinds of information, to which it has provided both the technical vocabulary and at least the initial conceptual framework. It would be impossible to understand the nature of information without grasping at least its main gist. This is the task of the present chapter.



## The mathematical theory of communication

MTC treats information as data communication, with the primary aim of devising efficient ways of encoding and transferring data.

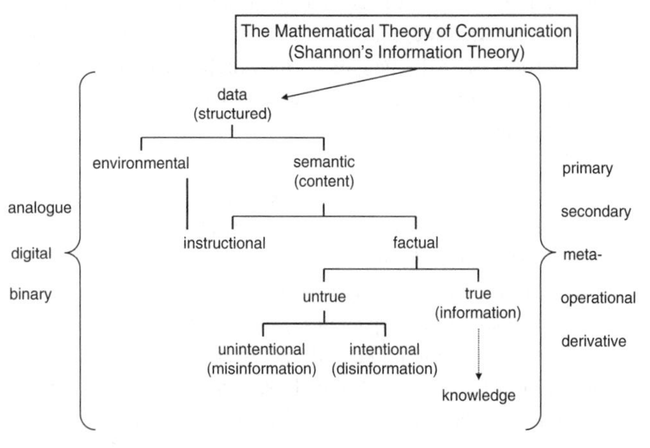


**7. The mathematical theory of communication (MTC)**

It has its origin in the field of electrical engineering, as the study of communication limits, and develops a quantitative approach to information.

To have an intuitive sense of the approach, let us return to our example. Recall John's telephone conversation with the mechanic. In Figure 8, John is the *informer*, the mechanic is the *informee*, 'the battery is flat' is the (semantic) message (the *informant*) sent by John, there is a coding and decoding procedure through a language (English), a channel of communication (the telephone system), and some possible noise (unwanted data received but not sent). Informer and informee share the same background knowledge about the collection of usable symbols (technically known as the *alphabet*; in the example, this is English).

MTC is concerned with the efficient use of the resources indicated in Figure 8. John's conversation with the mechanic is fairly realistic and hence more difficult to model than a simplified case. In order to introduce MTC, imagine instead a very boring device that can produce only one symbol. Edgar Alan Poe (1809–1849) wrote a short story in which a raven can answer only 'nevermore' to any question. Poe's raven is called a *unary device*. Imagine John rings the garage and his call is answered by Poe's raven. Even at this elementary level, Shannon's simple model of communication still

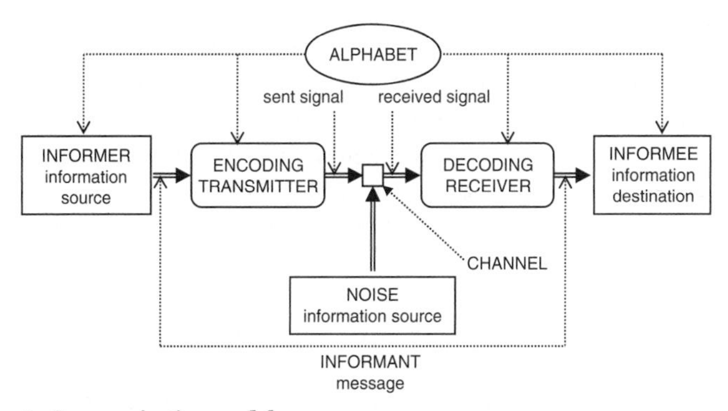


**8. Communication model**

applies. It is obvious that the raven (a unary device) produces zero amount of information. Simplifying, John already knows the outcome of the communication exchange: whatever he asks, the answer is always 'nevermore'. So his ignorance, expressed by his question, e.g. 'can I recharge the battery?', cannot be decreased. Whatever his informational state is, asking appropriate questions to the raven, e.g. 'will I be able to make the car start?', 'can you come to fix the car?', makes no difference. Note that, interestingly enough, this is the basis of Plato's famous argument, in the *Phaedrus*, against the value of semantic information provided by written texts:

> [Socrates]: Writing, Phaedrus, has this strange quality, and is very like painting; for the creatures of painting stand like living beings, but if one asks them a question, they preserve a solemn silence. And so it is with written words; you might think they spoke as if they had intelligence, but if you question them, wishing to know about their sayings, they always say only one and the same thing [they are unary devices, in our terminology]. And every word, when [275e] once it is written, is bandied about, alike among those who understand and those who have no interest in it, and it knows not to whom to speak or not to speak; when ill-treated or unjustly reviled it always needs its father to help it; for it has no power to protect or help itself.



As Plato well realizes, a unary source answers every question all the time with only one message, not with silence or message, since silence counts as a message, as we saw in Chapter 2. It follows that a completely silent source also qualifies as a unary source. And if silencing a source (censorship) may be a nasty way of making a source uninformative, it is well known that crying wolf (repeating always the same message, no matter what the circumstances are) is a classic case in which an informative source degrades to the role of an uninformative unary device.

Consider now a binary device that can produce two messages, like a fair coin $A$ with its two equiprobable symbols *heads* and *tails*, that is, $\{h, t\}$; or, as Matthew 5:37 suggests, 'Let your communication be

Yea, yea; Nay, nay: for whatsoever is more than these cometh of evil'. Before the coin is tossed, the informee (for example a computer) does not 'know' which symbol the device will actually produce: it is in a state of *data deficit* greater than zero. Shannon used the technical term 'uncertainty' to refer to such a data deficit. In a non-mathematical context, this can be misleading because of the strong psychological connotations of this term, so one may wish to avoid it. Recall that the informee can be a simple machine, and psychological or mental states are clearly irrelevant. Once the coin has been tossed, the system produces an amount of information that is a function of the possible outputs, in this case two equiprobable symbols, and equal to the data deficit that it removes. This is one bit of information. Let us now build a slightly more complex system, made of two fair coins *A* and *B*. The *AB* system can produce four results: <*h*, *h*>, <*h*, *t*>, <*t*, *h*>, <*t*, *t*>. It generates a data deficit of four units, each couple counting as a symbol <_, _>, in the source alphabet. In the *AB* system, the occurrence of each symbol <_, _> removes a higher data deficit than the occurrence of a symbol in the *A* system. In other words, each symbol provides more information by excluding more alternatives. Adding an extra coin would produce an eight units of data deficit, further increasing the amount of information carried by each symbol <_, _, _> in the *ABC* system, and so forth (see Table 5).



**Table 5. Examples of communication devices and their information power**

| Device | Alphabet | Bits of information per symbol |
|---|---|---|
| Poe's raven (unary) | 1 symbol | $\log(1) = 0$ |
| 1 coin (binary) | 2 equiprobable symbols | $\log(2) = 1$ |
| 2 coins | 4 equiprobable symbols | $\log(4) = 2$ |
| 1 die | 6 equiprobable symbols | $\log(6) = 2.58$ |
| 3 coins | 8 equiprobable symbols | $\log(8) = 3$ |

The basic idea is that information can be quantified in terms of decrease in data deficit (Shannon's 'uncertainty'). One coin produces one bit of information, two coins produce two bits, three coins three bits, and so forth. Unfortunately, real coins are always biased. To calculate how much information they really produce one must rely on the frequency of the occurrences of symbols in a finite series of tosses, or on their probabilities, if the tosses are supposed to go on indefinitely. Compared to a fair coin, a slightly biased coin must produce less than one bit of information, but still more than zero. The raven produced no information at all because the occurrence of a string of 'nevermore' was not *informative* (not *surprising*, to use Shannon's more intuitive, but psychologistic vocabulary), and that is because the *probability* of the occurrence of 'nevermore' was maximum, so completely predictable. Likewise, the amount of information produced by the biased coin depends on the average *informativeness* of the occurrence of *h* or *t*. The more likely one of the results is, the less surprised we will be in being told the result, the less informative the outcome is. When the coin is so biased to produce always the same symbol, it stops being informative at all, and behaves like the raven or the boy who cries wolf.



The quantitative approach just sketched plays a fundamental role in coding theory, hence in cryptography, and in data storage and transmission techniques. MTC is primarily a study of the properties of a channel of communication and of codes that can efficiently encipher data into recordable and transmittable signals. Two concepts that play a pivotal role both in communication analysis and in memory management are so important to deserve a brief explanation: *redundancy* and *noise*.

## Redundancy and noise

In real life, a good codification is modestly redundant. *Redundancy* refers to the difference between the physical representation of a message and the mathematical representation of the same message that uses no more bits than necessary. *Compression* procedures, such as those used to reduce the digital size of photographs, work by reducing data redundancy, but redundancy is not always a bad thing, for it can help to counteract *equivocation* (data sent but never received) and *noise*. A message + noise contains more data than the original message by itself, but the aim of a communication process is *fidelity*, the accurate transfer of the original message from sender to receiver, not data increase. We are more likely to reconstruct a message correctly at the end of the transmission if some degree of redundancy counterbalances the inevitable noise and equivocation introduced by the physical process of communication and the environment. Noise extends the informee's freedom of choice in selecting a message, but it is an undesirable freedom and some redundancy can help to limit it. That is why the manual of John's car includes both verbal explanations and pictures to convey (slightly redundantly) the same information.



## Some conceptual implications of the mathematical theory of communication

For the mathematical theory of communication (MTC), information is only a selection of one symbol from a set of possible symbols, so a simple way of grasping how MTC quantifies information is by considering the number of yes/no questions required to determine what the source is communicating. One question is sufficient to determine the output of a fair coin, which therefore is said to produce one bit of information. We have seen that a two-fair-coins system produces four ordered outputs: <*h*, *h*>, <*h*, *t*>, <*t*, *h*>, <*t*, *t*> and therefore requires at least two questions, each output containing two bits of information, and so on. This analysis clarifies two important points.

First, MTC is not a theory of information in the ordinary sense of the word. In MTC, information has an entirely technical meaning. For a start, according to MTC, two equiprobable 'yes' answers contain the same amount of information, no matter whether their corresponding questions are 'is the battery flat?' or 'would you marry me?'. If we knew that a device could send us, with equal probabilities, either this book or the whole *Encyclopedia Britannica*, by receiving one or the other we would receive very different amounts of bytes of data but actually only one bit of information in the MTC sense of the word. On 1 June 1944, the BBC broadcast a line from Verlaine's *Song of Autumn*: '*Les sanglots longs des violons de Autumne*'. This was a coded message containing less than one bit of information, an increasingly likely 'yes' to the question whether the D-Day invasion was imminent. The BBC then broadcast the second line '*Blessent mon coeur d'une longueur monotone*'. Another almost meaningless string of letters, but almost another bit of information, since it was the other long-expected 'yes' to the question whether the invasion was to take place immediately. German intelligence knew about the code, intercepted those messages, and even notified Berlin, but the high command failed to alert the Seventh Army Corps stationed in Normandy. Hitler had all the information in Shannon's sense of the word, but failed to understand (or believe in) the crucial importance of those two small bits of data. As for ourselves, we should not be surprised to conclude that the maximum amount of information, in the MTC sense of the word, is produced by a text where each character is equally distributed, that is by a perfectly random sequence. According to MTC, the classic monkey randomly pressing typewriter keys is indeed producing a lot of information.



Second, since MTC is a theory of information without meaning (not in the sense of meaningless, but in the sense of not yet meaningful), and since [information – meaning = data], 'mathematical theory of data communication' is a far more appropriate description of this branch of probability theory than 'information theory'. This is not a mere question of labels. Information, as semantic content (more on this shortly), can also be described as *data + queries*. Imagine a piece of information such as 'the earth has only one moon'. It is easy to polarize almost all its semantic content by transforming it into a [query + binary answer], such as [does the earth have only one moon? + yes]. Subtract the 'yes' – which is at most one bit of information – and you are left with all the semantic content, with all the indications of its truth or falsity removed. Semantic content is information not yet saturated by a correct answer. The datum 'yes' works as a key to unlock the information contained in the query. MTC studies the codification and transmission of information by treating it as data keys, that is, as the amount of detail in a signal or message or memory space necessary to saturate the informee's unsaturated information. As Weaver correctly remarked:

> the word information relates not so much to what you do say, as to what you could say. The mathematical theory of communication deals with the carriers of information, symbols and signals, not with information itself. That is, information is the measure of your freedom of choice when you select a message.

MTC deals with messages comprising uninterpreted symbols encoded in well-formed strings of signals. These are mere data that constitute, but are not yet, semantic information. So MTC is commonly described as a study of information at the *syntactic* level. And since computers are syntactical devices, this is why MTC can be applied so successfully in ICT.

## Entropy and randomness

Information in Shannon's sense is also known as *entropy*. It seems we owe this confusing label to John von Neumann (1903–1957), one of the most brilliant scientists of the 20th century, who recommended it to Shannon:

> You should call it entropy for two reasons: first, the function is already in use in thermodynamics under the same name; second, and more importantly, most people don't know what entropy really is, and if you use the word *entropy* in an argument you will win every time.

Von Neumann proved to be right on both accounts, unfortunately. Assuming the ideal case of a noiseless channel of communication, entropy is a measure of three equivalent quantities:

(a) the average amount of information per symbol produced by the informer, or
(b) the corresponding average amount of data deficit (Shannon's uncertainty) that the informee has before the inspection of the output of the informer, or
(c) the corresponding informational potentiality of the same source, that is, its *informational entropy.*

Entropy can equally indicate (a) or (b) because, by selecting a particular alphabet, the informer automatically creates a data deficit (uncertainty) in the informee, which then can be satisfied (resolved) in various degrees by the *informant.* Recall the game of questions and answers. If you use a single fair coin, I immediately find myself in a one bit deficit predicament: I do not know whether it is heads or tails, and I need one question to find out. Use two fair coins and my deficit doubles, as I need at least two questions, but use the raven, and my deficit becomes null. My empty glass (point (b) above) is an exact measure of your capacity to fill it (point (a) above). Of course, it makes sense to talk of information as quantified by entropy only if one can specify the probability distribution.

Regarding (c), MTC treats information like a physical quantity, such as mass or energy, and the closeness between its analysis of information and the formulation of the concept of entropy in statistical mechanics was already discussed by Shannon. The informational and the thermodynamic concept of entropy are related through the concepts of probability and *randomness*. 'Randomness' is better than 'disorder', since the former is a syntactical concept, whereas the latter has a strongly semantic value, that is, it is easily associated to interpretations, as I used to try to explain to my parents as a teenager. Entropy is a measure of the amount of 'mixedupness' in processes and systems bearing energy or information. It can also be seen as an indicator of reversibility: if there is no change of entropy then the process is reversible. A highly structured, perfectly organized message contains a lower degree of entropy or randomness, less information in the Shannon sense, and hence it causes a smaller data deficit, which can be close to zero (recall the raven). By contrast, the higher the potential randomness of the symbols in the alphabet, the more bits of information can be produced by the device. Entropy assumes its maximum value in the extreme case of uniform distribution, which is to say that a glass of water with a cube of ice contains less entropy than the glass of water once the cube has melted, and a biased coin has less entropy than a fair coin. In thermodynamics, the greater the entropy, the less available the energy is (see Chapter 5). This means that high entropy corresponds to high energy deficit, but so does entropy in MTC: higher values of entropy correspond to higher quantities of data deficit. Perhaps von Neumann was right after all.

Our exploration of the quantitative concepts of information is complete. MTC provides the foundation for a mathematical approach to the communication and processing of well-formed data. When these are meaningful, they constitute *semantic content* (see Chapter 2). When semantic content is also true, it qualifies as semantic information. This is the queen of all concepts discussed in this book and the next chapter is dedicated to it.

# Chapter 4
# Semantic information

Let us go back to John's conversation with the mechanic. The mathematical theory of communication (MTC) provides a detailed analysis of how the exchange of data over the phone works. As far as MTC is concerned, John and the mechanic might have been talking about the weather, or some problem with the car breaking system, or indeed anything else. This is so because MTC studies information as a probabilistic phenomenon. Its central question is whether and how much uninterpreted data can be encoded and transmitted efficiently by means of a given alphabet and through a given channel. MTC is not interested in the meaning, reference, relevance, reliability, usefulness, or interpretation of the information exchanged, but only in the level of detail and frequency in the uninterpreted data that constitute it. Thus, the difference between information in Shannon's sense and semantic information is comparable to the difference between a Newtonian description of the physical laws describing the dynamics of a tennis game and the description of the same game as a Wimbledon final by a commentator. The two are certainly related, the question is how closely. In this chapter, we shall look at the definition of semantic information, then explore several approaches that have sought to provide a satisfactory account of what it means for something to be semantically informative. We shall then consider

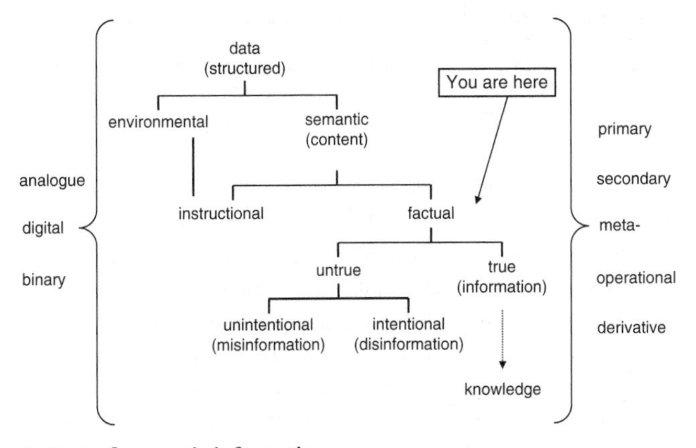


**9. Factual semantic information**



two significant problems affecting such approaches, the Bar-Hillel Carnap paradox and the scandal of deduction, and how they may be solved.

## Factual semantic information

Semantic content may be *instructional*, as when John is told over the phone how to use jumper cables to start his car's engine; or *factual*, as when John tells the mechanic that the battery is flat (see Chapter 2). But what is the difference between some semantic content and some semantic information, when they are both factual? Recall John's lie: he said to the mechanic that his wife had forgotten to switch off the car's lights, when in fact he had. Did John provide any information to the mechanic? Strictly speaking, he provided only a false 'story', that is, some semantic content about a plausible situation. In fact, he failed to inform the mechanic because the semantic content was not true. In formal terms, the definition ([Def]) of semantic information is:

**Table 6. The definition of factual semantic information**

| | |
|---|---|
| [DEF] | *p* qualifies as factual semantic information if and only if *p* is (constituted by) *well-formed*, *meaningful*, and *veridical data*. |

[DEF] captures the general consensus reached by the debate. According to it, factual semantic information is, strictly speaking, inherently *truth-constituted* and not a contingent *truth-bearer*, exactly like knowledge but unlike propositions or beliefs, for example, which are what they are independently of their truth values. Semantic information encapsulates truth, exactly as knowledge does: the mechanic fails to be informed and hence to know that John's wife forgot to switch off the car's lights because it is not true that she did, but he is informed and hence knows that the battery of John's car is flat because it is true that it is. So the difference between factual semantic content (see the definition GDI, Table 1, Chapter 2) and factual semantic information is that the latter needs to be true, whereas the former can also be false. Note that in [DEF] we speak of veridical rather than true data because strings or patterns of well-formed and meaningful data may constitute sentences in a natural language, but of course they can also generate formulae, maps, diagrams, videos, or other semiotic constructs in a variety of physical codes, and in these cases 'veridical' is to be preferred to 'true'.

[DEF] offers several advantages, three of which are worth highlighting in this context. First, it clarifies the fact that false information is not a genuine type of information. One speaks of false information not as one speaks of a false sentence, which is a sentence that happens to be false, but in the same way as one qualifies someone as a false friend, i.e. not a friend at all. It follows that when semantic content is false, this is a case of *misinformation*. If the source of misinformation is aware of its nature, as when John intentionally lied to the mechanic, one speaks of *disinformation*. Disinformation and misinformation are ethically censurable but may be successful in achieving their purpose: in our example, the mechanic was still able to provide John with the right advice, despite being disinformed by him about the exact cause of the problem. Likewise, information may still fail to be successful; just imagine John telling the mechanic that his car is merely out of order.

The second advantage is that [DEF] forges a robust and intuitive link between factual semantic information and knowledge. Knowledge encapsulates truth because it encapsulates semantic information, which, in turn, encapsulates truth, as in a three dolls matryoshka. Knowledge and information are members of the same conceptual family. What the former enjoys and the latter lacks, over and above their family resemblance, is the web of mutual relations that allow one part of it to account for another. Shatter that, and you are left with a pile of truths or a random list of bits of information that cannot help to make sense of the reality they seek to address. Build or reconstruct that network of relations, and information starts providing that overall view of the world which we associate with the best of our epistemic efforts. So once some information is available, knowledge can be built in terms of explanations or accounts that make sense of the available semantic information. John knows that the battery is flat not by merely guessing rightly, but because he connects into a correct account the visual information that the red light of the low-battery indicator is flashing, with the acoustic information that the engine is not making any noise, and with the overall impression that the car is not starting. In this sense, semantic information is the essential starting point of any scientific investigation.

A third advantage will be appreciable towards the end of this chapter, where [DEF] plays a crucial role in the solution of the so-called Bar-Hillel Carnap paradox. Before that, we need to understand what it means for something to convey the information that such and such is the case, that is, in what sense semantic information may be more or less informative, and whether this 'more or less' may be amenable to rigorous quantification.

## The analysis of informativeness

Approaches to the informativeness of semantic information differ from MTC in two main respects. First, they seek to give an account of information as *semantic* content, investigating questions such as 'how can something count as information? and why?', 'how can something carry information about something else?', 'how can semantic information be generated and flow?', 'how is information related to error, truth and knowledge?', 'when is information useful?'. Second, approaches to semantic information also seek to connect it to other relevant concepts of information and more complex forms of epistemic and mental phenomena, in order to understand what it means for something, such as a message, to be informative. For instance, we may attempt to ground factual semantic information in environmental information. This approach is also known as the *naturalization of information*.



Analyses of factual semantic information tend to rely on propositions, such as 'Paris is the capital of France', 'Water is $H_2O$', or 'the car's battery is flat'. How relevant is MTC to similar analyses? In the past, some research programmes tried to elaborate information theories *alternative* to MTC, with the aim of incorporating the semantic dimension. Nowadays, most researchers agree that MTC provides a rigorous constraint to any further theorizing on all the semantic and pragmatic aspects of information. The disagreement concerns the crucial issue of the *strength* of the constraint.

At one extreme of the spectrum, a theory of factual semantic information is supposed to be *very strongly* constrained, perhaps even overdetermined, by MTC, somewhat as mechanical engineering is by Newtonian physics. Weaver's optimistic interpretation of Shannon's work, encountered in the Introduction, is a typical example.

At the other extreme, a theory of factual semantic information is supposed to be *only weakly* constrained, perhaps even completely underdetermined, by MTC, somewhat as tennis is constrained by Newtonian physics, that is, in the most uninteresting, inconsequential, and hence disregardable sense.

The emergence of MTC in the 1950s generated some initial enthusiasm that gradually cooled down in the following decades. Historically, theories of factual semantic information have moved from 'very strongly constrained' to 'only weakly constrained'. Recently, we find positions that appreciate MTC only for what it can provide in terms of a robust and well-developed statistical theory of correlations between states of different systems (the sender and the receiver) according to their probabilities.

Although the analysis of semantic information has become increasingly autonomous from MTC, two important connections have remained stable between MTC and even the most recent accounts: the communication model, explained in Chapter 3; and the so-called 'Inverse Relationship Principle' (IRP).



The communication model has remained virtually unchallenged, even if nowadays theoretical accounts are more likely to consider, as basic cases, multiagent and distributed systems interacting in parallel, rather than individual agents related by simple, sequential channels of communication. In this respect, our philosophy of information has become less Cartesian and more 'social'.

IRP refers to the inverse relation between the probability of $p$ – where $p$ may be a proposition, a sentence of a given language, an event, a situation, or a possible world – and the amount of semantic information carried by $p$. IRP states that information goes hand in hand with unpredictability (Shannon's surprise factor). Recall that Poe's raven, as a unary source, provides no information because its answers are entirely predictable. Likewise, a biased coin provides increasingly less information the more likely one of its outcomes is, to the point that if it had two identical sides, say heads, the probability of heads would be one while the informativeness of being told that it is heads would be zero. Karl Popper (1902–1994) is often credited as the first to have advocated IRP explicitly. However, systematic attempts to develop a formal calculus involving it were made only after Shannon's breakthrough. MTC defines information in terms of probability. Along similar lines, the *probabilistic approach* to semantic information defines the information in *p* in terms of the inverse relation between information and the probability of *p*. This approach was initially suggested by Yehoshua Bar-Hillel (1915–1975) and Rudolf Carnap (1891–1970). Several approaches have refined their work in various ways. However, they all share IRP as a basic tenet, and for this reason they all encounter two classic problems, known as the 'scandal of deduction' and the Bar-Hillel–Carnap paradox.



## The scandal of deduction

Following IRP, the more probable or possible *p* is, the less informative it is. So, if the mechanic tells John that a new battery will be available sometime in the future, this is less informative than if he tells him that it will be available in less than a month, since the latter message excludes more possibilities. This seems plausible, but consider what happens when the probability of *p* is highest, that is, when $P(p) = 1$. In this case, *p* is equivalent to a tautology, that is, something that is always true. Tautologies are well known for being non-informative. John would be receiving data but no *semantic* information if he were told that 'a new battery will or will not become available in the future'. Again, this seems very reasonable. However, in classical logic, a conclusion Q is deducible from a finite set of premises $P_1, \ldots, P_n$ if and only if the conditional [$P_1$ and $P_2$, and... $P_n$ imply Q] is a tautology. Accordingly, since tautologies carry no information at all, no logical inference can yield an increase of information, so logical deductions, which can be analysed in terms of tautological processes, also fail to provide any information. Indeed, by identifying the semantic information carried by a sentence with the set of all possible worlds or circumstances it excludes, it can be recognized that, in any valid deduction, the information carried by the conclusion must be already contained in the information carried by the (conjunction of) the premises. This is what is often meant by saying that tautologies and inferences are 'analytical'. But then logic and mathematics would be utterly uninformative. This counterintuitive conclusion is known as 'the scandal of deduction'. Here is how the philosopher and logician Jaakko Hintikka (born 1929) describes it:

> C. D. Broad has called the unsolved problems concerning induction a scandal of philosophy. It seems to me that in addition to this scandal of induction there is an equally disquieting scandal of deduction. Its urgency can be brought home to each of us by any clever freshman who asks, upon being told that deductive reasoning is 'tautological' or 'analytical' and that logical truths have no 'empirical content' and cannot be used to make 'factual assertions': in what other sense, then, does deductive reasoning give us new information? Is it not perfectly obvious there is some such sense, for what point would there otherwise be to logic and mathematics?



There have been many attempts to solve the problem. Some refer to the psychological nature of logical informativeness. According to this view, the role of logical reasoning is that of helping us to bring out the full informational content of sentences, so that one can clearly see that the conclusion is indeed contained in the premises by simple inspection. It is as if the premises of a logical deduction were like compressed springs: they do not generate new information but merely store it and then release it again once they return to their original shape, namely once the deduction is fully laid down to include the conclusions. Logic and mathematics yield an increase of information but only for limited minds like ours, which cannot see how the conclusion is already implicit in the premises. This approach is unsatisfactory, since it fails to explain why, if the conclusion of a deductive argument is always 'contained' in the premises, deductive reasoning is generally perceived as highly valuable for scientific purposes. If all theorems are 'contained' in the axioms of a theory, mathematical discoveries would be impossible. Moreover, interesting theorems are usually very hard to prove in terms of computational resources. Other approaches have shown that classic, logico-mathematical deductions are informative because proof of their validity essentially requires the (temporary) introduction of 'virtual information', which is assumed, used, and then discharged, thus leaving no trace at the end of the process, but hugely contributing to its success. An elementary example will help to clarify the point.



Suppose John has the following information 'the battery of the car is flat (call this case P) *and/or* the car's electric system is out of order (call this case Q)'. Let us abbreviate *and/or* as $\vee$, meaning that either P or Q or both could be the case. The mechanic tells John that *if* P is the case, *then* someone from the garage will have to come to fix the problem (call this scenario S) and that *if* Q is the case, *then* again S is the case. Let us abbreviate *if… then* with the symbol $\rightarrow$. John's updated information now looks like this:

1) $P \vee Q$
2) $P \rightarrow S$
3) $Q \rightarrow S$

Note that (1)–(3) is all the actual information that John has. John does not have the information that the battery of the car is flat, nor does he have the information that the car's electric system is out of order, but only that at least one but possibly both problems have occurred. However, John is good at logic, so he tries to calculate what is going to happen by making some *assumptions*, that is, he steps out of the available space of information, represented by (1)–(3), and pretends to have more information than he really has, strictly speaking. His reasoning (see Figure 10) is: '*suppose* P is the

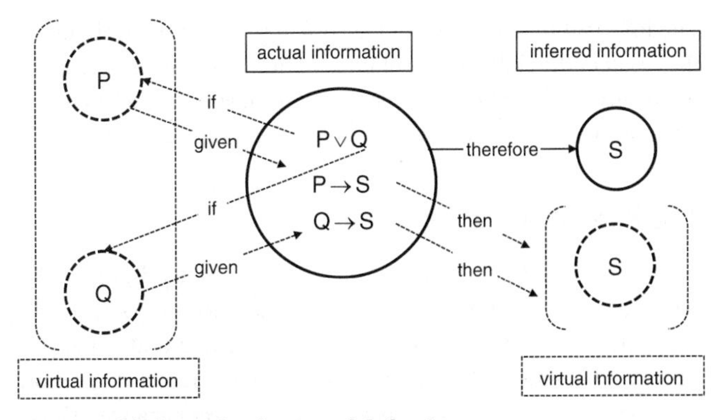


**10. Virtual information in natural deduction**

case. From (2) it already follows that S. But *suppose* Q is the case, then from (3) it already follows that S. But then, I really do not need to *suppose* either P or Q by themselves, since I do have them packed together in (1), so from (1), (2), and (3) I can infer that S: someone from the garage will have to come to fix the problem.' John has just used what in natural deduction systems is known as the '∨ elimination rule'. He started with a disjunction (1), then treated each disjunct in turn as an assumption, then tried to prove that the assumption (together with other available premises) entails the conclusion. Having succeeded in showing that either disjunct suffices to entail the conclusion, he discharged the assumptions and asserted the conclusion. Although the process is simple and very obvious, it is also clear that John quietly stepped out of the space of information that he actually had, moved in a space of virtual information, made it do quite a lot of essential work, and then stepped back into the original space of information that he did have, and obtained his conclusion. If one does not pay a lot of attention, the magic trick is almost invisible. But it is exactly the stepping in and out of the space of available information that makes it possible for deductions to be at the same time formally valid and yet informative.

The informational richness of logico-mathematical deductions is the result of the skilful usage of informational resources that are by no means contained in the premises, but must nevertheless be taken into consideration in order to obtain the conclusion.

## The Bar-Hillel–Carnap paradox

Let us return to IRP. The less probable or possible *p* is, the more informative it is. If John is told that the car's electric system is out of order, this is more informative than if he is told that either the battery is flat and/or the car's electric system is out of order, simply because the former case is satisfied by fewer circumstances. Once again, this seems reasonable. But if we keep making *p* less and less likely, we reach a point when the probability of *p* is actually zero, that is, *p* is impossible or equivalent to a contradiction, but, according to IRP, this is when *p* should be maximally informative. John would be receiving the highest amount of semantic information if he were told that the car's battery is and is not flat (at the same time and in the same sense). This other counterintuitive conclusion has been called the Bar-Hillel–Carnap paradox (because the two philosophers were among the first to make explicit the counterintuitive idea that contradictions are highly informative).

Since its formulation, the problem has been recognized as an unfortunate, yet perfectly correct and logically inevitable consequence of any quantitative *theory of weakly semantic information*. 'Weakly' because truth values play no role in it. As a consequence, the problem has often been either ignored or tolerated as the price of an otherwise valuable approach. A straightforward way of avoiding the paradox, however, is by adopting a semantically stronger approach, according to which factual semantic information encapsulates truth. Once again, the technicalities can be skipped in favour of the simple idea. The reader may recall that one of the advantages of [DEF] was that it could play a crucial role in the solution of the Bar-Hillel–Carnap paradox. It is now easy to see why: if something qualifies as factual semantic information only when it satisfies the truthfulness condition, contradictions and indeed falsehoods are excluded *a priori*. The quantity of semantic information in *p* can then be calculated in terms of distance of *p* from the situation *w* that *p* is supposed to address. Imagine there will be exactly three guests for dinner tonight. This is our situation *w*. Imagine John is cooking the meal and he is told that:

A. there will or will not be some guests for dinner tonight; or
B. there will be some guests tonight; or
C. there will be three guests tonight; or
D. there will and will not be some guests tonight.

The *degree of informativeness* of A is zero because, as a tautology, A applies both to *w* and to its negation. B performs better, while C has the maximum degree of informativeness because, as a fully accurate, precise, and contingent truth, it 'zeros in' on its target *w*. And since D is false (it is a contradiction), it does not qualify as semantic information at all, just mere content (see Figure 11 in the next chapter). Generally, the more distant the information is from its target, the larger the number of situations to which it applies, and the lower its degree of informativeness becomes. A tautology is an instance of true information that is most 'distant' from the world. A contradiction is an instance of misinformation that is equally distant from the world. Of course, sometimes one may prefer an instance of misinformation – e.g. being told that there will be four guests, when in fact there will be only three – than an instance of semantic information that is too vacuous – e.g. being told that there will be fewer than a hundred guests tonight.

# Chapter 5
# Physical information

So far, information has been analysed from a mathematical and a semantic perspective. However, as anyone who has suffered the heat of a laptop knows too well, information is also a physical phenomenon. Storing and processing data is energy-consuming, and this is why data centres have started raising serious ecological problems (see the Epilogue). In Chapter 3, the concept of entropy was applied both in information theory and in thermodynamics. So the time has come to consider what physics (as a theory of phenomena) and metaphysics (as a theory of what might lie behind those phenomena) have to say about the nature of information. The two perspectives are not incompatible and may be complementary.

Note that Figure 11 is not meant to suggest that semantic information is not physical. A road sign, indicating the maximum speed allowed by law, is a very physical implementation of some semantic information. What the map is meant to indicate is that, in this chapter, the semantic aspects of structured data will be disregarded in favour of their physical features as a natural phenomenon occurring in the environment.

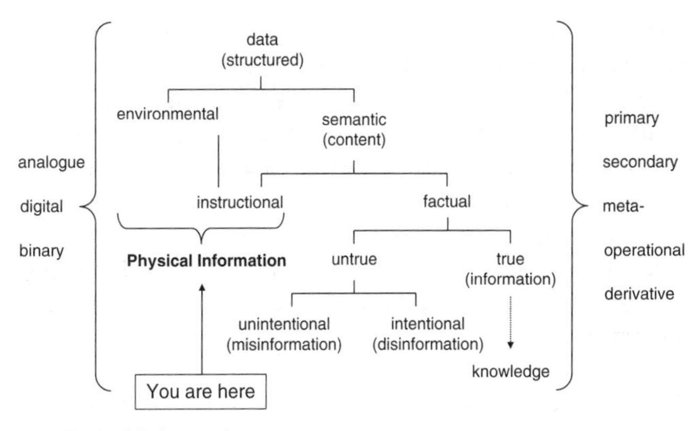


**11. Physical information**



## Maxwell's demon

Thermodynamics studies the conversion of energy from one kind (e.g. kinetic, mechanical, chemical, or electrical) to another, the direction in which energy flows, and the availability of energy to do work. It is therefore the scientific area that most contributed to the Industrial Revolution, since it provided the foundation for the efficient working of engines, such as the steam and the internal-combustion engines, which in turn made possible mechanical transports and the automatic manufacturing of goods. As the science of energy processes, thermodynamics has always enjoyed a double relation with the dynamics of information. On the one hand, information processes seem inevitably physical, hence based on energy transformations and therefore subject to thermodynamic laws. On the other hand, the design, improvement, and efficient management of thermodynamic processes themselves may heavily depend upon intelligent ways of handling information processes. Consider our example. Any information exchange that John enjoys with the world (the red light flashing, his telephone call, his conversation with his neighbour, etc.), requires energy transformations in the systems involved (his body, the car, etc.), which are ultimately subject to thermodynamic laws. At the same time, quite a lot of energy could have been saved if the whole thermodynamic process, which caused the flat battery, had been prevented by an audible signal, warning John that the lights were still on when he was exiting the car. Thermodynamics and information theory are often allies sharing one goal: the most efficient use of their resources, energy, and information.

Their potential degree of efficiency might seem to be boundless: the better we can manage information (e.g., extract or process more information with the same or less energy), the better we can manage energy (extract more, recycle more, use less or better), which can then be used to improve information processes further, and so forth. Is there a limit to such a virtuous cycle? In Chapter 3, I mentioned the fact that the mathematical theory of communication provides a boundary to how much one can improve the information flow physically. Unfortunately, thermodynamics now provides two further constraints on how far one can improve physical processes informationally.

The first law of thermodynamics concerns the conservation of energy. It states that the change in the internal energy of a closed thermodynamic system is equal to the sum of the amount of heat energy supplied to the system and the work done on the system. In other words, it establishes that the total amount of energy in an isolated system remains constant: energy can be transformed, but it can neither be created nor destroyed. So, no matter how smart and efficient our information handling might become, it is impossible to devise a perpetual motion machine, that is, a mechanism that, once started, will continue in motion indefinitely, without requiring any further input of energy. The most ingenious system ever designed will still require some energy input. The 'green' challenge is to use information more and more intelligently in order to reduce that energy input to ever lower targets, while keeping or increasing the output.

A sceptic may accept the previous limit and yet object that a perpetual motion machine is impossible because we have imagined using information only to build it from without, in terms of a very ingenious design. We should consider the possibility of putting some information device inside it, in order to regulate it from within. ICTs have made this trivial: 'smart' applications are everywhere and require no leap of sci-fi imagination. The answer to such an objection is twofold.

First, the second law of thermodynamics makes such 'smart' perpetual motion machines *physically* impossible. We have already encountered the concept of entropy in terms of 'mixedupness'. In thermodynamics, this is equivalent to a measure of the unavailability of a system's energy to do work, given the fact that available energy to do work requires some inequality (i.e. non-mixedupness) in states. According to the second law, the total entropy of any isolated thermodynamic system tends to increase over time, approaching a maximum value. Heat cannot by itself flow from a cooler to a hotter object. Reversing such process would be like observing a glass of lukewarm water with some ice in it spontaneously freeze, instead of observing the ice slowly melt: a miracle.

The second answer is much trickier. For the second law states that entropy *tends* to increase, so one may wonder whether it would be *logically* impossible (i.e. a contradiction in terms, like happily married bachelors) to observe the water molecules freeze, using the previous example. In other words, a fair coin will not land all the time on only one side, and this is a fact, but that possibility is not excluded by the laws of logic alone, it is not a contradiction. So, could one imagine some logically possible mechanism whereby entropy could be defeated just in theory, although still never in practice? Enter Maxwell's demon.

James Clerk Maxwell (1831–1879), the father of classical electromagnetic theory, devised his thought experiment in order to clarify what he saw as the statistical nature of the second law. In his *Theory of Heat*, he invited the reader to imagine the following scenario (see Figure 12). A container, filled with some gas, is divided into two parts, A and B. In the division, there is a microscopic hole, and a being, later to be known as Maxwell's demon, can open it or close it by means of a trapdoor. He monitors the molecules bouncing around at different speeds. When they approach the trapdoor, he opens it to allow faster-than-average molecules to move from A to B, and slower-than-average molecules to move from B to A. Eventually, he sorts out all particles into slower (A) and faster (B), thus providing an exception to the second law: mixedupness has decreased without any supply of energy.



It was soon realized that Maxwell's demon is an information device, which monitors and computes the trajectories of the particles. If it were theoretically possible, we would have identified a logically possible way of using information to defy physical entropy, by generating work from a system at energy costs lower than those required by the second law (recall that average

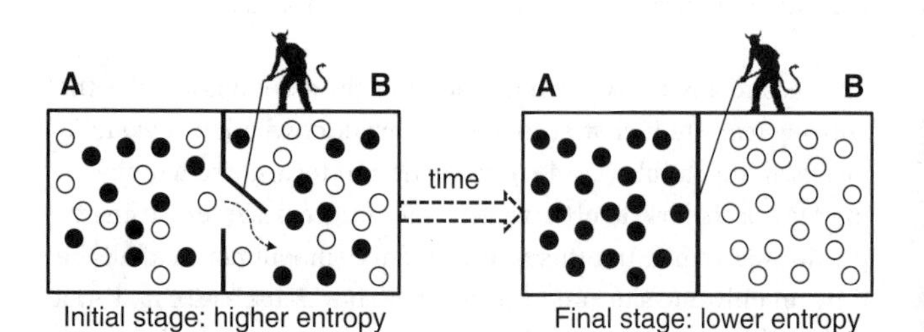


**12. Maxwell's demon**

molecular speed corresponds to temperature, which will have decreased in A and increased in B, making possible some work). Yet the second law of thermodynamics seems incontrovertible, so where is the trick? In Maxwell's original version, the demon needs to operate the trapdoor, which requires some energy, yet this feature of the thought experiment can be taken care of by designing it slightly differently (with sliding doors, mechanical devices, springs and so forth). As two great physicists, Leó Szilárd (1898–1964) and Léon Brillouin (1889–1969), realized, the real trick is in the information processes carried on by the demon. Any information collection, such as monitoring the location and speed of the particles, requires energy. Imagine, for example, the demon using a light beam to 'see' where the particles are: the photons that bounce off the particles to indicate their positions will have been produced by a source of energy. And even if further improvements in the design of the system could overcome this particular limit, there is a final constraint. Once information has been collected, the demon must perform some information processing, such as calculating exactly when to operate the trapdoor, in order to work effectively and hence decrease the entropy of the system. But computation uses memory – the demon needs to store information first, in order to manipulate it afterwards – no matter how efficiently. Therefore, as our demon keeps operating, the entropy will decrease, yet its memory storage will increase. Two computer scientists then finally managed to exorcise the demon. First, Rolf W. Landauer (1927–1999) argued that any logically irreversible manipulation of information causes the release of a specific amount of heat and thus generates a corresponding increase in entropy in the environment. Then Charles H. Bennett (born 1943) proved that most computations can be done reversibly, so that energy costs can be regained and entropy may not be increased, but that there is one computational operation which is necessarily irreversible, namely memory erasure (see Chapter 2). So the demon will need energy to erase its memory and this energy is what pays the entropy bill of the system under the counter, so to speak.

The conclusion is that information is a physical phenomenon, subject to the laws of thermodynamics. Or so it seemed until recently. For our story has an open ending. Landauer's principle is not a law and it has been challenged, in recent years, for actually presupposing, rather than supporting, the second law of thermodynamics. Moreover, one may argue that it is logically possible (although not physically feasible, but then this is why the demon is a thought experiment not a blueprint) that the demon may not need to erase its memory. If no information is ever erased, all his other computations may in principle be achieved in ways that are thermodynamically reversible, requiring no release of heat and hence no increase in entropy. We would end with a draw: proving that the system does not run for free, but failing to make the demon pay for the energy bill. Bloated with ever increasing amounts of recorded data, the demon would represent an ever-expanding space of memories.



Our sceptic may have a last objection. Maxwell's demon can see and manipulate single particles. If it were a quantum computer, couldn't this provide a solution to the amount of informational resources needed to defeat the second law of thermodynamics? The short answer is no; the long answer requires a new section.

## Quantum information

Binary data are encoded, stored, and processed by allowing each bit to be only in one fully determined, definite state at a time. The coins in Chapter 3 were classical, Newtonian systems, in which a conventional bit is either 0 or 1, on or off, heads or tails, and so forth, and can represent only a single value. However, quantum states of atomic particles have a peculiar nature. They can be used to store data in a definable but still undetermined quantum superposition of two states at the same time. Metaphorically, the reader may wish to refer to those pictures, made famous by Maurits Cornelis Escher, that contain two equally valid but incompatible interpretations at the same time. For example, one can see alternatively but not simultaneously the face of an old woman or the profile of a young one. The result of such a superposition of states is known as a qubit (quantum bit). A qubit is actually in both the 0-state and the 1-state *simultaneously*, although possibly to different extents. It is a vacillating unit of information, and it is only once its state is observed or measured that it invariably collapses to either 0 or 1. This physical phenomenon of superposition of states is ordinary in nature but strongly counterintuitive to our common sense, since it is difficult to grasp how a qubit could be in two opposite states simultaneously.

A quantum computer (QC) handles qubits, and this is why, if it could be built, it would be extraordinarily powerful. Suppose our simple computer works with only three coins. Each coin can be either 0 or 1, and there is a total of 8 possible combinations, that is, $2^3$, where 2 is the number of states and 3 the number of coins. This is known as a 3-bit register. A classic computer, using a 3-bit register can only operate sequentially on one of its 8 possible states at a time. To prepare each state of a register of 8 states a classic computer needs 8 operations. Take now a 3-qubit register QC. With some simplifications, we can 'load' the quantum register to represent all $2^3$ states simultaneously, because now $n$ elementary operations can generate a state containing $2^n$ possible states. A single QC can then perform 8 operations at once, sifting through all the qubit patterns simultaneously. This is known as *quantum parallelism*: having the whole matrix of 8 states in front of itself in one operation, a QC can explore all possible solutions of the problem in one step. The larger its register, the more exponentially powerful a QC becomes, and a QC with a register of 64 qubits could outsmart any network of supercomputers.

Quantum computers, if physically implemented, would then represent new types of information systems, alternative to our present computers, based on simple Newtonian physics. Their greater computational power would force us to rethink the nature and limits of computational complexity. Not only would QCs make present applications in cryptography, based on factoring difficulties, obsolete, they would also provide new means to generate absolutely secure cryptosystems and, more generally, transform into trivial operations statistical computations that are of extraordinary complexity.



Unfortunately, despite some successes with very elementary systems, the difficulties in building an actual QC that could replace your laptop may turn out to be insurmountable. Some physics of information is very hard to bend to our needs and qubits are exceedingly fragile artefacts. As for our sceptic, even a quantum version of Maxwell's demon would still incur into the constraints discussed in the previous section. And the computational limits of a QC are the same as those of a classic computer: it can compute recursive functions that are in principle computable by our classic machines (effective computation). It is exponentially more efficient than an ordinary computer, in that it can do much more in much less time. But this is a quantitative not a qualitative difference, which concerns the physical resources used to deal with information. Classical computing is based on the fact that space resources (location, memory, stability of physical states, etc.) are not a major problem, but time is. Quantum computing deals with the time-related difficulties of classical computing (some information processing just takes far too much time) by means of a shift. The relation between computational time (how many steps) and space (how much memory) is inverted: time becomes less problematic than space by transforming quantum phenomena of superposition, which are short-lasting and uncontrollable at a microscopic level, into quantum phenomena that are sufficiently long-lasting and controllable at a macroscopic level to enable computational processes to be implemented. Quantum computers will become a commodity only if this shift will become empirically feasible. Physicists could then use quantum information as a powerful means to model and investigate quantum mechanics hypotheses and other phenomena that are computationally too demanding for our present technology. Indeed, according to some researchers, they might discover that reality (the 'it') in itself is made of information (the 'bit'), the topic of our last section.

## It from bit

In Chapter 2, we saw that data in the wild were described as 'fractures in the continuum' or lacks of uniformity in the fabric of reality. Although there can be no information without data, data might not require a material implementation. The principle 'no information without data representation' is often interpreted materialistically, as advocating the impossibility of physically disembodied information, through the equation 'representation = physical implementation'. This is an inevitable assumption in the physics of information systems, where one must necessarily take into account the physical properties and limits of the data carriers and processes. But the principle in itself does not specify whether, ultimately, the occurrence of digital or analogue states necessarily requires a *material* implementation of the data in question. Several philosophers have accepted the principle while defending the possibility that the universe might ultimately be non-material, or based on a non-material source. Indeed, the classic debate on the ultimate nature of reality could be reconstructed in terms of the possible interpretations of that principle.

All this explains why the physics of information is consistent with two slogans, this time popular among scientists, both favourable to the proto-physical nature of information. The first is by Norbert Wiener (1894–1964), the father of cybernetics: 'information is information, not matter or energy. No materialism which does not admit this can survive at the present day.' The other is by John Archibald Wheeler (1911–2008), a very eminent physicist, who coined the expression 'it from bit' to indicate that the ultimate nature of physical reality, the 'it', is informational, comes from the 'bit'. In both cases, physics ends up endorsing an information-based description of nature. The universe is fundamentally composed of data, understood as *dedomena*, patterns or fields of differences, instead of matter or energy, with material objects as a complex secondary manifestation.

This informational metaphysics may, but does not have to, endorse a more controversial view of the physical universe as a gigantic digital computer, according to which dynamic processes are some kind of transitions in computational states. The distinction may seem subtle, but it is crucial. Imagine describing the stomach as if it were a computer (with inputs, processing stages, and output) versus holding that the stomach actually is a computer. Whether the physical universe might be effectively and adequately modelled digitally and computationally is a different question from whether the ultimate nature of the physical universe might be actually digital and computational in itself. The first is an empirico-mathematical question that, so far, remains unsettled. The second is a metaphysical question that should probably be answered in the negative, at least according to the majority of physicists and philosophers. One reason is because the models proposed in digital physics are not easily reconcilable with our current understanding of the universe. For example, Seth Lloyd estimates that the physical universe, understood as a computational system, could have performed $10^{120}$ operations on $10^{90}$ bits ($10^{120}$ bits including gravitational degrees of freedom) since the Big Bang. The problem is that, if this were true, the universe would 'run out of memory' because, as Philip Ball has remarked:

> To simulate the Universe in every detail since time began, the computer would have to have $10^{90}$ bits – binary digits, or devices

capable of storing a 1 or a 0 – and it would have to perform $10^{120}$ manipulations of those bits. Unfortunately there are probably only around $10^{80}$ elementary particles in the Universe.

Moreover, if the world were a computer, this would imply the total predictability of its developments and the resuscitation of another demon, that of Laplace.

Pierre-Simon Laplace (1749–1827), one of the founding fathers of mathematical astronomy and statistics, suggested that if a hypothetical being (known as Laplace's demon) could have all the necessary information about the precise location and momentum of every atom in the universe, he could then use Newton's laws to calculate the entire history of the universe. This extreme form of determinism was still popular in the 19th century, but in the 20th century was undermined by the ostensibly probabilistic nature of quantum phenomena. Science has moved from being based on necessity and laws to being based on probability and constraints. Nowadays, the most accepted view in physics is that particles behave indeterministically and follow the uncertainty principle. To the best of our knowledge – that is, at least according to the Copenhagen interpretation of quantum mechanics, which is the most widely accepted among physicists – computational determinism is not an option, Laplace's demon is a ghost, and digital physics shares its fate.

A digital reinterpretation of contemporary physics may still be possible in theory, but a metaphysics based on information-theoretic grounds seems to offer a more promising approach. Following Wiener and Wheeler, one might interpret reality as constituted by information, that is, by mind-independent, structural entities that are cohering clusters of data, understood as concrete, relational points of lack of uniformity. Such structural reality allows or invites certain constructs and resist or impede some others, depending on the interactions with, and the nature of, the information systems inhabiting it, e.g. *inforgs* like us. If an informational approach to the nature of reality is satisfactory, what could it tells us about the nature of life? And how do biological organisms cope with patterns of data? This is the topic of the next chapter.

# Chapter 6
# Biological information

Biological information may have so many meanings, and be used for such a variety of purposes that it can easily become too generic and lose most of its explanatory value. To see why, let us go back to John's interactions with his environment (see Figure 13).

As a living organism, John has a genetic code. As an agent, he inputs information from the environment through perceptual processes (e.g. he sees the red light flashing), elaborates such environmental information through internal neurophysiological processes (e.g. he realizes that if the red light is flashing then the battery must be flat), and outputs semantic information into the environment through communication processes (e.g. by talking to his neighbour). Each stage may qualify as a case of biological information; the input-elaboration-output processes are not as clear cut as I just presented them, but are strictly intertwined with each other; they may be studied by more than once discipline (philosophy of mind, neuroscience, psychology, physiology, epistemology, information theory, and so forth) with its own technical vocabulary; and several concepts of information are sprinkled all over the place. The mess is almost inevitable and often irrevocable. So, in order to avoid getting lost, this chapter will analyse only two aspects of the already very simplified and

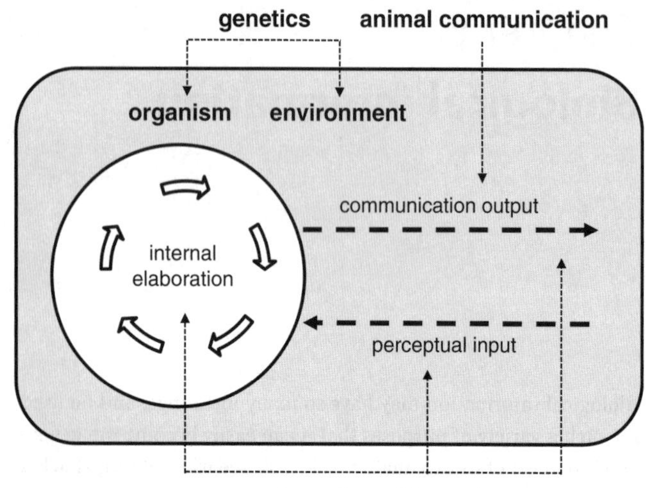


**13. Biological information**



schematic picture sketched in Figure 13: the nature of genetic information (John as an organism) and how information is used in neuroscience, which, for lack of a better term, I shall call neural information (John as a brain). The input and the output phase were discussed in the previous chapters.

Before beginning our explorations, two conceptual distinctions will come in handy. First, it will be useful to recall that there are three main ways of talking about information:

(a) Information *as* reality, e.g. patterns, fingerprints, tree rings;
(b) Information *for* reality, e.g. commands, algorithms, recipes;
(c) Information *about* reality, i.e. with an epistemic value, e.g. train tables, maps, entries in an encyclopaedia.

Something may count as information in more than one sense, depending on the context. For example, a person's iris may be an instance of information *as* reality (the pattern of the membrane in the eye), which provides information *for* reality (e.g. as a biometric means to open a door by verifying the identity of a person), or *about* reality (e.g. the identity of the person). But it is crucial to be clear about what sense of information is being used in each case: (a) *physical*, (b) *instructional*, or (c) *semantic*. Unfortunately, biological information is often used ambiguously in all three senses at the same time.

The second distinction is equally conceptual but might be phrased more easily linguistically, in terms of two different uses of 'biological' or 'genetic':

A) *attributive*: biological (genetic) information is information *about* biological (genetic) facts.

P) *predicative*: biological information is information *whose nature* is biological (genetic) in itself.



Consider the following examples: medical information is information about medical facts (attributive use), not information that has curative properties; digital information is not information about something digital, but information that is in itself of digital nature (predicative use); and military information can be both information about something military (attributive) and of military nature in itself (predicative). When talking about biological or genetic information, the attributive sense is common and uncontroversial. In bioinformatics, for example, a database may contain medical records and genealogical or genetic data about a whole population. Nobody disagrees about the existence of this kind of biological or genetic information. It is the predicative sense that is more contentious. Are biological or genetic processes or elements intrinsically informational in themselves? If biological or genetic phenomena count as informational *predicatively*, is this just a matter of modelling, that is, may be seen as being informational? If they really are informational, what sort of informational stuff are they? And what kind of informational concepts are needed in order to understand their nature? The next section should help to provide some answers.

## Genetic information

Genetics is the branch of biology that studies the structures and processes involved in the heredity and variation of the genetic material and observable traits (phenotypes) of living organisms. Heredity and variations have been exploited by humanity since antiquity, for example to breed animals. But it was only in the 19th century that Gregor Mendel (1822–1884), the founder of genetics, showed that phenotypes are passed on, from one generation to the next, through what were later called genes. In 1944, in a brilliant book based on a series of lectures, entitled *What Is Life?*, the physics Nobel laureate Erwin Schrödinger (1887–1961) outlined the idea of how genetic information might be stored. He explicitly drew a comparison with the Morse alphabet. In 1953, James Watson (born 1928) and Francis Crick (1916–2004) published their molecular model for the structure of DNA, the famous double helix, one of the icons of contemporary science. Crick explicitly acknowledged his intellectual debt to Schrödinger's model. In 1962, Watson, Crick, and Maurice Wilkins (1916–2004) were jointly awarded the Nobel Prize for Physiology or Medicine 'for their discoveries concerning the molecular structure of nucleic acids and its significance for information transfer in living material'. Information had become one of the foundational ideas of genetics. Let us see why.

John has 23 pairs of chromosomes in the cells of his body (sperm, eggs, and red blood cells are the only exceptions), one of each pair from his mother and the other from his father. Each of his chromosomes consists of proteins and DNA (deoxyribonucleic acid, see Figure 14), the molecule that contains the genetic code for all life forms apart from some viruses. DNA is made up of chemical units called nucleotides. Each nucleotide contains one of four bases (adenine = A, guanine = G, cytosine = C, and thymine = T), one phosphate molecule, and the sugar molecule deoxyribose. A gene is a segment of a DNA molecule that contains information for making functional molecules such as RNA (ribonucleic acid) and proteins, which perform the chemical reactions in the organism.

John's genetic code is stored on one of the two long, twisted strands of his DNA, as a linear, non-overlapping sequence of A, G, C, and T. This is the 'alphabet' used to write the 'code words', known as *codons*. Each codon is a unique combination of three letters, which are eventually interpreted as a single amino acid in a chain. Since the letters are four and the positions they can occupy are three, there are $4^3 = 64$ possible combinations or codons. One of these codons acts as start signal and begins all the sequences that code for amino acid chains. Three of these codons act as stop signals and indicate that the message is complete. All the other sequences code for specific amino acids.

In order to obtain a protein from a gene, two very complex and still not entirely understood processes are needed, known as *transcription* and *translation*. Through transcription, or RNA synthesis, DNA nucleotide sequence information is copied into RNA sequence information. The resulting, complementary nucleotide RNA strand is called messenger RNA (mRNA), because it carries a genetic message from the DNA to the system of the cell that synthesizes the protein. Through translation, or protein initial biosynthesis, mRNA (the output of the transcription process) is decoded to produce proteins. The mRNA sequence works as a template to produce a chain of amino acids that are then assembled into protein. Once it is correctly produced, the protein starts working and the genetic trait associated to it is generated. Occasionally, there may be an accidental error (change, duplication, gap) in the reproduction of the DNA sequence of a gene. This casual genetic mutation may affect the production of proteins. It may be harmless (no effect), harmful (negative effect), or advantageous (positive effect). In the latter case, it generates

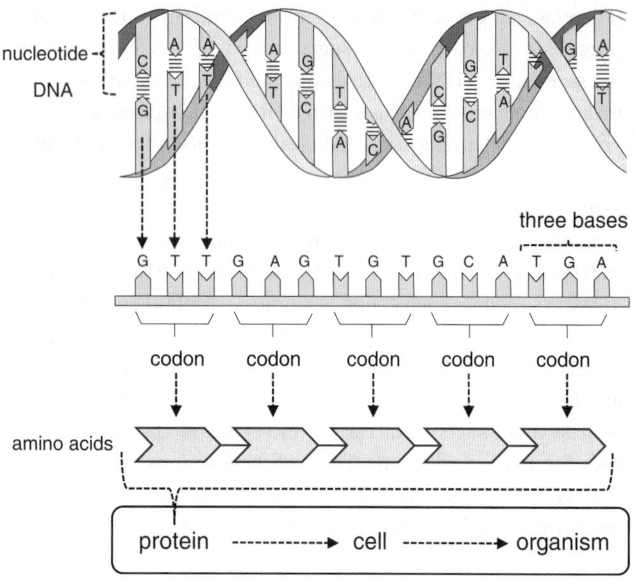




**14. DNA and the genetic code**

new versions of proteins that give a survival advantage to the organisms in question. In the long run, such random genetic mutations make possible the evolution of new forms of life.

Despite this very general outline, it is already obvious that the role played in genetics by informational concepts is crucial. The question is what biological information could be (in the predicative use of biological), given that there are so many different concepts of information.

We saw in Chapter 3 that information in Shannon's sense provides the necessary ground to understand other kinds of information. So, if biological information is indeed a kind of information, to be told that it must satisfy the causal constraints and physical correlations identified by the mathematical theory of information is to be told too little. Some researchers, perceiving the need of a more substantive explanation, have opted for a semantic interpretation of biological information. This, however, seems to be an overreaction. In the precise sense in which one may speak of semantic information, genetic information can hardly count as an instance of it. It simply lacks all its typical features, including meaningfulness, intentionality, aboutness, and veridicality. DNA contains the genetic code, precisely in the sense that it physically contains the genes which code for the development of the phenotypes. So DNA does contain genetic information, like a CD may contain some software. But the genetic code or, better, the genes, are the information itself. Genes do not *send* information, in the sense in which a radio sends a signal. They work more or less successfully and, like a recipe for a cake, may only partly guarantee the end result, since the environment plays a crucial role. Genes do not *contain* information, like envelopes or emails do, nor do they *describe* it, like a blueprint; they are more like performatives: 'I promise to come at 8 pm' does not describe or contain a promise, it does something, namely it effects the promise itself through the uttered words. Genes do not *carry* information, as a pigeon may carry a message, no more than a key carries the information to open the door. They do not *encode* instructions, as a string of lines and dots may encode a message in Morse alphabet. True, genes are often said to be the *bearers* of information, or to *carry instructions* for the development and functioning of organisms, and so forth, but this way of speaking says more about us than about genetics. We regularly talk about our current computers as if they were intelligent – when we know they are not – and we tend to attribute semantic features to genetic structures and processes, which of course are biochemical and not intentional at all. The 'code' vocabulary should not be taken too literally, as if genes were information in a *semantic-descriptive* sense, lest we run the risk of obfuscating our understanding of genetics. Rather, genes are instructions, and instructions are a type of predicative and effective/procedural information, like recipes, algorithms, and commands. So genes are dynamic procedural structures that, together with other indispensable environmental factors, contribute to control and guide the development of organisms. This is a perfectly respectable sense in which biological information is indeed a kind of information. Dynamic procedural structures are a special type of informational entities, those that are in themselves instructions, programs, or imperatives.

The previous interpretation is compatible with, and complements, the mathematical theory of information, but is less demanding than a semantic interpretation. It has the advantage of explaining how genes achieve what they achieve, since it interprets them as instructions that require the full collaboration of the relevant components of the organism and of its environment to be carried on successfully. And it can clarify the informational approach to the genetic code in computational terms that are much better understood and completely non-intentional, namely by drawing a comparison with imperative programming (also known as procedural programming) in computer science. In imperative programming, statements change a program state and programs are a sequence of commands for the computer to perform. Each step (each base) is an instruction, and the physical environment holds the state that is modified by the instruction. The relation between instructions (genes, imperative programs, recipes) and the outcome is still functional, causal, and based on laws, but no semantics needs to be invoked, exactly in the same sense in which no semantics plays any role in the way computer hardware is designed to execute machine code, which is written in the imperative style and native to the computer. So, with a slogan, in the genetic code, the medium (the genes) is the message. Biological information, in the predicative sense of the world, is procedural: it is information *for* something, not *about* something. Genetic information can now be placed on our map (see Figure 15).

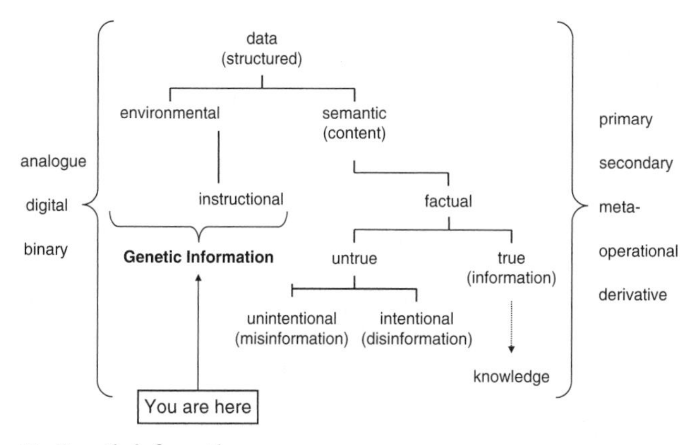


**15. Genetic information**



A final comment before switching to neural information. Of course, genes play a crucial role in explaining not only the development of individual organisms but also the inheritance of phenotypes across generations. Therefore, informational approaches have been adopted both in genetics and in evolutionary biology and even at the higher level of biological anthropology. *Memes* (alleged units or elements of cultural ideas, symbols, or practices), for example, have been postulated as cultural analogues to genes, which are transmitted from one mind to another through communication and imitable phenomena, self-replicating and responding to selective pressures. In similar contexts, however, there is the danger that the concept of biological information might lose its useful and concrete procedural sense, and silently acquire an increasing semantic sense. This shift towards the right-hand side of our map might be suggestive, but it should be considered to have a heuristic value at best, only as a way for us to solve specific problems or discover new features to the topics investigated. It is more metaphorical than empirical, and it can be hardly explanatory in terms of physical correlations and interacting mechanisms.

## Neural information

Without genetic modifications, John would have never developed. He and almost all other animals (sponges are among the few exceptions) belong to the so-called *bilateria*. These are organisms with a bilaterally symmetric body shape. Fossil evidence shows that bilateria probably evolved from a common ancestor around 550 million years ago. Much to John's disappointment, that ancestor was a humble tube worm. Luckily for him, it was a rather special one. It is still unclear exactly when and how bilateria evolved a nervous system and how this further evolved in different groups of organisms. But, at a crucial point, John's ancestor acquired a segmented body, with a nerve cord that had an enlargement, called a ganglion, for each body segment, and a rather larger ganglion at the end of the body, called the brain. The ultimate anti-entropic weapon had been born. Biological life is a constant struggle against thermodynamic entropy. A living system is any anti-entropic informational entity, i.e. an informational object capable of instantiating procedural interactions (it embodies information-processing operations) in order to maintain its existence and/or reproduce itself (metabolism). Even single-celled organisms extract and react to information from their environment to survive. But it is only with the evolution of a sophisticated nervous system that can gather, store, process, and communicate vast amounts of information and use it successfully, that the implementation and control of a larger variety of anti-entropic behaviours became possible. After millions of years of evolution, nervous systems can now be found in many multicellular animals. They can greatly differ in size, shape, and complexity between species. So, for our present purposes, we shall simply concentrate on John as an anti-entropic informational agent, and sketch his information-processing abilities.

From a bioinformational perspective, John's nervous system is a network that manages information about his environment and himself, causing bodily actions and reactions that benefit him as an organism, promote his welfare, and increase his chances of survival and reproduction. The building elements of such a network are *neurons* and *glia*, roughly in a ratio of ten glia to one neuron. Glia are specialized cells that surround neurons, providing them with insulation, nutrients, and some specialized chemicals involved in electrochemical signalling. Neurons are specialized cells, which integrate, receive, and send different types of electrochemical signals, with different patterns of activity. Although neurons come in a very large variety of shapes, sizes, and functional capacities, they can be categorized according to their sensory-motor function. *Sensory neurons* receive sensory information from the periphery of the body. *Motor neurons* control muscle movements. *Interneurons* are interposed between sensory and motor neurons. Most neurons share a common structure, and Figure 16 provides a very simplified and schematic representation of some of the major components of a typical neuron.



The cell body is called the *soma*. The soma has a *nucleus* and tree-like structures, known as *dendrites*, where signals from other neurons are received and integrated. Specialized protrusions (*spines*) can be present to optimize the reception of signals. Typically, the spines or dendrites contain receptor molecules on the membrane surface, which react with a chemical transmitter substance – a neurotransmitter – released by the adjacent neuron.

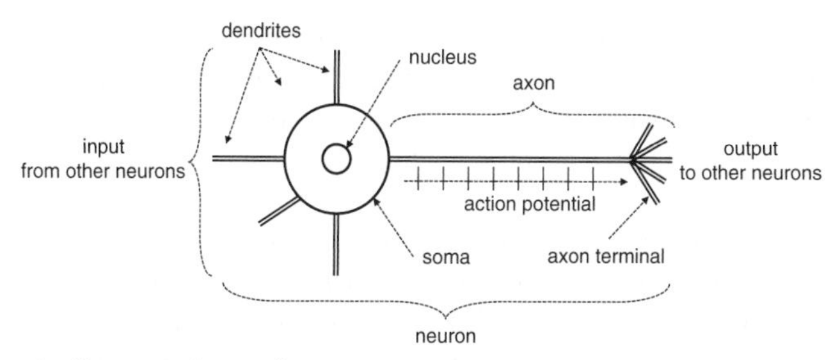


**16. Abstract scheme of a neuron**

The binding of a neurotransmitter with a receptor causes the opening of a pore (*channel*) in the membrane through which charged particles (*ions*) can flow. The result of current flow through the channel can facilitate (excitatory) or inhibit (inhibitory) the probability that the receiving neuron will become activated and send its own electrochemical signal to other cells. The *axon* is a long, slender tube that relays the electrochemical signal. Axons are the primary transmission lines of the nervous system, and as bundles they help to make up nerves. They have microscopic diameters, typically about 1 micrometre across (one millionth of a metre, or 1/1,000 of a millimetre; consider that a strand of human hair is about 100 micrometre), but they may be very long. The longest axons in the human body are those of the sciatic nerve, which run from the base of the spine to the big toe of each foot.

The electrochemical signal carried by the axon is called the *action potential* (AP). The AP is a Boolean (all-or-none) electrochemical signal transmitted down the axon into its terminals, where it causes neurotransmitter release. Physically, the AP is a brief (regenerative) spike of about 100 millivolts (a millivolt is 1/1,000 of a volt). It lasts about 1–3 milliseconds and travels along the axon at speeds of 1–100 meters per second. Information about the strength of activation in a neuron, and consequently the information that it carries, is transmitted by the frequency (rate) of action potentials, since the magnitude and duration of the AP are not variable enough to code changes in activation. Some neurons emit action potentials constantly, at rates of 10–100 per second, usually in irregular temporal patterns; other neurons are quiet most of the time, but occasionally emit a burst of action potentials.

Terminal buttons are the endings of the axons, where the arriving action potential is converted into the release of the neuron's transmitter substance. Most axons branch extensively, and individual neurons can have thousands of terminals. Typically, terminals contain packets filled with the neurotransmitter molecule. Voltage-sensitive receptors are activated by the action potential, and result in the opening of channels in the terminal membrane. A cascade of biochemical events ensues, which results in the release of neurotransmitter substance. The synapse is the junction between two neurons, where the electrochemical signal is exchanged.

There are several chemicals that transmit excitatory and inhibitory signals at synapses. The effect that a neurotransmitter has upon a neuron depends on the receptor molecules that it activates. In some cases, the same neurotransmitter can either be excitatory or inhibitory, or can have very fast or very long-lasting effects. Drugs such as caffeine can mimic or interfere with brain activity by facilitating or inhibiting the neurotransmitter action. Several amino acids have been suggested to serve as neurotransmitters. The most common neurotransmitters in the mammalian brain are glutamate and gamma-aminobutyric acid (GABA). Given their simplicity and ubiquity, and their presence in simpler organisms, they may have been some of the earliest transmitters to evolve.



Neurons work by transforming chemical signals into electrical impulses and vice versa. So the nervous system is a complex network that process data electrochemically. The architecture of the network is usually organized into a peripheral and central nervous system. The peripheral system consists of sensory neurons and the neurons that connect them to the spinal cord and the brain. The latter makes up the central nervous system. So the peripheral system is the interface between John's body and the external world's physical data flow (lights, sounds, smells, pressures, etc.) and coordinates his movements, including his physiological states and functions. Sensory neurons respond to external data input (the physical stimuli) by generating and propagating internal data (the signals) to the central nervous system, which then processes and communicates the elaborated data (signals) back to the bodily system. At the centre of the network architecture there is another complex network, the brain. John's brain consists of approximately 100 billion neurons, each linked with up to 10,000 synaptic connections.

Despite this quick sketch, it is obvious why the nervous system, and the brain in particular, are studied from an informational perspective. On the one hand, neuroinformatics develops and applies computational tools, methods, models, approaches, databases, and so forth, in order to analyse and integrate experimental data and to improve existing theories about the structure and function of the brain. On the other hand, computational neuroscience investigates the information-theoretical and computational nature of biologically realistic neurons and neural networks, their physiology and dynamics. Thus, computer science and ICTs have provided extraordinary means to observe and record the brain (neuroimaging), such as electroencephalography (EEG, the recording of electrical activity along the scalp caused by the firing of neurons within the brain) and functional magnetic resonance imaging (fMRI, the measurement of blood-related dynamic responses connected to neural activity in the brain). Yet the brain is still a continent largely unexplored. One of the great informational puzzles is how physical signals, transduced by the nervous system, give rise to high-level, semantic information. When John sees the red light flashing, there is a chain of data-processing events that begins with an electromagnetic radiation in the environment, in the wavelength range of roughly 625–740 nanometres (one-billionth of a metre or one-millionth of a millimetre; red consists mainly of the longest wavelengths of light discernible by the human eye) and ends with John's awareness that there is a red light flashing in front of him probably meaning that the battery is flat. Some segments of this extraordinary journey are known, but large parts of it are still mysterious. Of course, there is nothing magic in it, but this is no insurance against the fact that the ultimate explanation may one day be astonishing.

An organism tends to act upon the world in a mediated way. It actively converts (sensory) data into information and then constructively processes this information to manage its interactions with the world. All this involves the elaboration of intermediary, internal constructs, which are stored, transformed, manipulated, and communicated over variable lengths of time, from short-term memory to over a lifetime. In humans, it involves the unique capacity to gather, store, and retrieve, exchange, integrate, and update, use and indeed misuse semantic information acquired by other people, including past generations. It is this social and economic sphere of information that will be explored in the next chapter.

# Chapter 7
# Economic information

In Oliver Stone's film *Wall Street* (1987), the main character, Gordon Gekko (Michael Douglas), declares that 'the most valuable commodity I know of is information'. He was probably right. Information has always had great value, and whoever has owned it has usually been keen on protecting it. This is why, for example, there are legal systems regulating intellectual property. Intellectual property rights concern artistic and commercial creations of the mind, and hence the relevant kinds of information and intangible assets. Copyrights, patents, industrial design rights, trade secrets, and trademarks are meant to provide an economic incentive to their beneficiaries to develop and share their information through a sort of temporary monopoly. Similarly, in many countries it is illegal to trade the securities of a corporation (e.g. bonds) on the basis of some privileged access to that corporation's non-public information, typically obtained while working for it (this is why it is referred to as insider trading). Military information is another good example. Julius Caesar (100 BC–44 BC) was so aware of the value of information that he devised one of the first and most popular encryption techniques, known as a Caesar cipher, to communicate with his generals. It consisted in replacing each letter in the message by a letter shifted some fixed number of positions down the alphabet. For example, with a shift of 4,

A would be replaced by E, B would become F, and so forth. Likewise, our computers are partly the outcome of the work done by Turing at Bletchley Park, Britain's code-breaking centre, to decipher German communications during the Second World War, and the internet developed during the Cold War to ensure that the US Air Force might still be able to share vital information even after a nuclear attack.

Clearly, when we talk about the economic value of information, the information in question is *semantic*. Although it is mathematically constrained and physically implemented – e.g. as a telephone call, an email, an oral message, a radio signal, a chemical formula, a Web page, or a map – it is the meaning conveyed by the information that is of value to the agents involved, who assume it to be correct or veridical (see Figure 17).



Economic value may be assigned to information according to the *price* it would bring in an open and competitive market (neoclassical economics). This is basically how much the agent interested in acquiring it would be ready to pay for it. Or economic value might be assigned in terms of the *amount of resources*, such

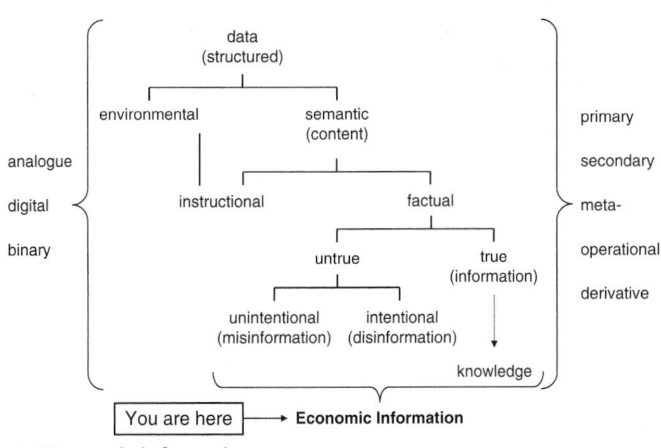


**17. Economic information**

as time, discomfort, or labour, that it would save to its holder (classical economics). This is basically how much benefit or lack of detriment the agent holding it would enjoy. The two interpretations are related. The economic value of information is the expected utility that results in the willingness to pay a corresponding price for it. In both cases, the information in question must have some features that are value-adding and value-preserving, such as timeliness, relevance, and updateness. Nobody pays for yesterday's newspaper or the wrong kind of information. Such features go under the general term of *information quality*.

When it is treated as a commodity, information has three main properties that differentiate it from other ordinary goods, such as cars or loaves of bread. First, it is *non-rivalrous*: John holding (consuming) the information that the battery is flat does not prevent the mechanic from holding (consuming) the same information at the same time. This is impossible with a loaf of bread. Secondly, by default, information tends to be *non-excludable*. Some information – such as intellectual properties, non-public and sensitive data, or military secrets – is often protected, but this requires a positive effort precisely because, normally, exclusion is not a natural property of information, which tends to be easily disclosed and shareable. By contrast, if John's neighbour lends his jump leads to him, he cannot use them at the same time. Finally, once some information is available, the cost of its reproduction tends to be negligible (*zero marginal cost*). This is of course not the case with many goods such as a loaf of bread. For all these reasons, information may be sometimes seen as a *public good*, a view which in turn justifies the creation of public libraries or projects such as *Wikipedia*, which are freely accessible to anyone.

Information has economic value because of its usefulness: it allows agents to take courses of actions (consider options, avoid errors, make choices, take rational decisions, and so forth) that normally yield higher expected payoffs (*expected utility*) than the agents would obtain in the absence of such information. The payoff may be understood, biologically, in terms of the informational organism's higher chances to withstand thermodynamic entropy to its own advantage. In economics, it is the sum of the *utilities* associated with all possible outcomes of a course of action, weighted by the probability that each outcome will occur, where 'utilities' refers to a measure of the relative satisfaction from, or desirability of, an outcome, e.g. the consumption of a good or service. The benefits brought about by information need to be understood contextually because the agents exchanging information could be not only human individuals, but also biological agents, social groups, artificial agents (such as software programs or industrial robots), or synthetic agents (such as a corporation or a tank), which comprise agents of all kinds.

In Chapter 1, we saw how human society has come to depend, for its proper functioning and growth, on the management and exploitation of information processes. Unsurprisingly, in recent years the scientific study of economic information has bloomed. In 2001, George Akerlof (born 1940), Michael Spence (born 1943), and Joseph E. Stiglitz (born 1943) were awarded what is known as the Nobel Prize in Economics 'for their analyses of markets with asymmetric information'. Indeed, information-theoretical approaches to economic topics have become so popular and pervasive that one may be forgiven for mistaking economics for a branch of information science. In the rest of this chapter, we will look at some essential ways in which economic information is used. For the sake of simplicity, and following current trends, the presentation will be framed in game-theoretic terms. But instead of presenting a standard analysis of types of games first, we will focus on the concepts of information and then see how they are used.

## Complete information

Game theory is the formal study of strategic situations and interactions (*games*) among agents (*players*, not necessarily human), who are fully rational (they always maximize their payoffs, without any concern for the other players), aware of each other, and aware that their decisions are mutually dependent and affect the resulting payoffs. Generally speaking, a game is described by four elements:

(a) its players, how many and who they are;
(b) each player's strategies, what they may rationally decide to do given the known circumstances (a strategy is a complete plan of action specifying a feasible action for every move the player might have to make);
(c) the resulting payoffs from each outcome, what they will gain by their moves; and
(d) the sequence (timing or order) of the actual moves or states, if the game is *sequential* (see below), basically in what position the player is at a certain stage of the game.



One of game theory's main goals is to identify the sort of stable situations (*equilibria*) in which the game players have adopted strategies that they are unlikely to change, even if, from a sort of God's eye perspective, they may not be rationally optimal. There are many kinds of game and hence forms of equilibrium. One way of classifying them is by checking how much game-relevant information the players enjoy, that is, who has what kind of access to (a)–(d).

A game is said to be based on *complete information* when all the players have information about (a), (b), and (c). Another way of defining complete information is in terms of common knowledge: each player knows that each player knows that... each player knows all other players, their strategies, and the corresponding payoffs for each player. Typical examples include the rock-paper-scissors game and the prisoner's dilemma. There is no need to describe the former but the latter is sufficiently complex to deserve some explanation.

We owe the logical structure of the prisoner's dilemma to the Cold War. In 1950, RAND Corporation (Research ANd Development, a non-profit think tank initially formed to provide research and analysis to the US armed forces) was interested in game theory because of its possible applications to global nuclear strategy. Merrill Flood (born 1912) and Melvin Dresher (1911–1992) were both working at RAND and they devised a game-theoretic model of cooperation and conflict, which later Albert Tucker (1905–1995) reshaped and christened as the 'prisoner's dilemma'. Here it is.

Two suspects, A and B, are arrested, but with insufficient evidence for a full conviction. So, once they are separated, each of them is offered the following deal. If one testifies against (*defects* from) the other, and the other remains silent (*cooperates*), the defector will not be charged but the cooperator will receive the full ten-year sentence. If both cooperate, each will receive a one-year sentence for a minor charge. If both defect, each will receive only half of the full sentence, five years. A and B must choose to defect from, or cooperate with, each other. Note that neither A nor B can know what the other will do. This is why this classic version of the prisoner's dilemma is a *simultaneous game*, exactly like the rock-paper-scissors game: it is not a matter of timing (rock-paper-scissors is also a *synchronic* game, with both players showing their hands at the same time), but of lack of information about the other player's (planned) move or state. What should each prisoner do?

The rational choice is for each prisoner to defect (five years in prison), despite the fact that each prisoner's individual payoff would be greater if they both cooperated (one year in prison). This may seem strange but, no matter what the other prisoner decides to do, each of them always gains a greater payoff by defecting. Since cooperating is *strictly dominated* by defecting, that is, since in any situation defecting is more beneficial than cooperating, defecting is the rational decision to take (Table 7). This sort of equilibrium qualifies as a Pareto-suboptimal solution (named after the economist Vilfredo Pareto, 1848–1923) because there could be a feasible change (known as Pareto improvement) to a situation in which no player would be worse off and at least one player would be better off. Unlike the other three outcomes, the case in which both prisoners defect can also be described as a *Nash equilibrium*: it is the only outcome in which each player is doing the best he can, given the available information about the other player's actions.

Nash equilibria are crucial features in game theory, as they represent situations in which no player's position can be improved by selecting any other available strategy while all the other players are also playing their best option and not changing their strategies. They are named after John Nash (born 1928), who, in 1994, shared the Nobel Prize in Economics with Reinhard Selten (born 1930) and John Harsanyi (1920–2000) for their foundational work on game theory.



Complete information makes simultaneous games interesting. Without such a condition, the players would be unable to predict the effects of their actions on the other players' behaviour. It is also a fundamental assumption behind the theoretical model of an efficient, perfectly competitive market, in which economic agents, e.g. buyers and sellers, consumers and firms, are supposed to hold all the necessary information to make optimal decisions. It is,

**Table 7. The normal form of a typical prisoner's dilemma. The matrix represents the players A and B, their strategies (columns and rows), and their payoffs (values in bold for A and underlined for B)**

| | | **Prisoner A** | |
|---|---|---|---|
| | | **Defect** | **Collaborate** |
| | Defect | 5 **5** | 0 **10** |
| Prisoner B | Collaborate | 10 **0** | 1 **1** |

however, a very strong assumption. Many games are based on *incomplete information*, with at least one player lacking information about at least one of the features (a)–(c). An interesting class of incomplete information games is based on the concept of *asymmetric information.*

## Asymmetric information

Suppose we treat the interactions between John and his insurer, called Mark, as a game. We know that John is very absent-minded (he tends to forget to switch off the lights of his car) and not entirely trustworthy (he tends to lie and likes to blame his wife for his mistakes). Mark, however, does not have all this information about John. So this is a case of asymmetric information: one player has relevant information that the other player misses. Mark is underinformed, and this can lead to two well-known types of problems: *moral hazard* and *adverse selection.*

An adverse selection scenario is one in which an absent-minded player like John is more likely to buy an insurance for his car battery because the underinformed player, like Mark, cannot adjust his response to him (e.g. by negotiating a higher premium) due to his lack of information (this is the relevant point here; Mark might also be bound by legal reasons even if he had enough information).

A moral hazard scenario is one in which, once John has had the battery of his car insured, he behaves even less carefully, e.g. by leaving the lights on and the iPod re-charging, because Mark, the underinformed player, does not have enough information about his behaviour (or does not have the legal power to use such information; again, the point of interest here is the informational one).

As the examples show, the two problems can easily be combined. Because of a known asymmetry in information, underinformed players tend to over-react. Mark will ask a higher premium from every customer because some of them will be like John. There arises a need for 'good' players to inform the underinformed players about themselves (indicate their 'types') and thus counterbalance the asymmetric relation. We have already encountered Spence and Stiglitz. Each of them developed an influential analysis of how asymmetric information might be overcome: *signalling* and *screening*, respectively.

Signalling may be described in terms of *derivative information* (see Chapter 2): the informed player provides reliable information which derivatively indicates the player's type to the underinformed player. Since signalling has been hugely influential in the literature on contract theory, here is the textbook example, slightly adapted.

When I first arrived in Oxford, I could not quite understand why so many very smart students would study Philosophy and Theology, running the obvious risk of being unemployed. Who needed platoons of philosophical theologians? I had not understood Spence's theory of signalling. Employers will hire, or pay higher wages to, applicants with better skills. However, they are underinformed about the actual skills of the applicants, all of whom will of course claim to have very high skills. In this case of asymmetric information (hiring is an investment decision under uncertainty), potential employees, our philosophical theologians, signal their types (high skills) to their potential, underinformed employers by acquiring a degree from a prestigious institution. This is costly, not just financially, but also in terms of competition, efforts, required skills, and so on. But for cost signalling to work, it is irrelevant what topics they have studied (primary information), as long as the fact of having obtained the costly degree sends the right signal (derivative information).

Screening can be seen as the converse of signalling. Instead of having informed players signalling their information, underinformed players induce informed players to disclose their information by offering a menu of options, e.g. different possible contracts, such that the choices of the informed agents reveal their information. For example, Mark might offer to John different combinations of premium or discounts to insure his car battery such that John's profile as a risky customer will become apparent.

One way in which underinformed players may overcome their disadvantage is by trying to reconstruct all the steps that led to the situation with which they are dealing. This is why a bank will normally interview a customer applying for a mortgage. When the game makes such information available by default, it is called perfect information.



## Perfect information

If players have total access to (d), that is, to the history of all the moves that have taken place in the game or the states in which the players are, they enjoy *perfect information*. Tic-tac-toe and chess are two examples of perfect-information games. They well illustrate a more formal definition of perfect information as an information *singleton* (a set with exactly one element). An information set establishes all the possible moves that could have taken place in the game, according to a particular player and given that player's information state. So, in a perfect-information game, every information set contains only one member (it is a singleton), namely the description of the point reached by the game at that stage. In tic-tac-toe, this will be a specific configuration of X and O on the $3 \times 3$ grid; in chess, this will be a specific configuration of all the playing pieces on the board. If the points were more than one (two configurations of the grid or of the board), the player would be uncertain about the history of the game, not knowing in which of the two situations the game is, and hence would not have perfect information.

Since complete information concerns features (a)–(c) of a game (players, strategies, and payoffs), while perfect information concerns only feature (d) (moves or states), clearly there can be games with complete and perfect information, such as chess; with only complete but not perfect information, as we saw in the previous section; and with only perfect information but no complete information. This may happen when players in the same game are actually playing a 'different' game, thus missing some information about feature (b) and/or (c). An example would be John playing chess with his daughter Jill but having a higher payoff for losing rather than winning the game as long as he lets her win without her noticing it.

Perfect information is an interesting feature of some sequential games. Games are said to be *sequential* when there is some predefined order according to which players move, and at least some players have information about the moves of players who preceded them. The presence of a sequence of moves is insufficient without some access to them, for in that case the game is effectively simultaneous and the difference in time has no strategic import. If all players have information about all the previous moves or states of all the other players then the sequential game is one of perfect information. Maxwell's demon and Laplace's demon (Chapter 6) may be described as complete- and perfect-information single-player games. If only some players have perfect information, then we shall see below that the sequential game is one of imperfect information. Examples in this case include Scrabble, a game in which each player is not informed about what tiles are held by another player, and poker, for the same reason.

In sequential games, agents with incomplete or imperfect information are lacking something precious, either some information about features (a)–(c) or some information about feature (d) of the game. Incomplete-information games are also known as Bayesian games (see next section). In a Bayesian game,

Nature, that is, a source of randomness and uncertainty, is introduced as a player. Nature's role is twofold: it assigns a random variable to each player which can take values of types (e.g. player A could be of type x, y, or z), and associates some probabilities with those types. The type of a player determines that player's payoff function and the probability associated with the type is the probability that that player (for whom the type is specified) is that type. This uncertainty means that at least one player is not informed about the type of another player and the corresponding payoff function. So players have some initial beliefs about the type of each player but need to revise them during the game, on the basis of the new moves. By making Nature's moves (which are unknown) the source of the lack of information about the type and hence payoffs of at least one of the players, one can transform incomplete-information games into imperfect-information games. One may then find the Nash equilibria for the imperfect-information games and then generalize them into Bayes-Nash equilibria for incomplete-information games.

Whenever information is incomplete or imperfect, there is a general need to be able to gain as much as possible of the missing information – either about the players (types, strategies, or payoffs) or the history of the game – by 'retrodicting' (predicting backwards) from the information that one does hold, the information that one misses. This process of reverse inference is done through Bayesian reasoning.

## Bayesian information

As a branch of probability theory, Bayesianism has many applications and could have been introduced in other chapters. It is discussed here because it helps us to understand how underinformed agents might revise or upgrade their information in dynamic contexts in which their courses of action need to be refined as further information becomes available.

Let us begin by considering a simple example. John's daughter, Jill, receives many emails, and only a few of them (say 2%) are infected by some software virus. She uses a rather reliable antivirus software, which is successful 95% of the time, that is, it provides only 5% false positives. The software does not erase her potentially infected emails, but removes them to a special quarantine folder, which Jill can check. Jill would like to know how often she should check the folder for good emails. The question she is implicitly asking is: 'what is the probability that *A* (= the email was infected), given the fact that *B* (= the email was blocked by the antivirus and placed in the quarantine folder) when, on average, 2% of my emails are actually infected and my antivirus is successful 95% of the time?'. Jill has just identified a way of acquiring (learning) the missing piece of information that will help her to adopt the right strategy: if the chance that some emails in the quarantine folder might not be infected is very low, she will check it only occasionally. How could she obtain such a missing piece of information? The answer is by using a Bayesian approach.



Thomas Bayes (1702–1761) was a Presbyterian minister and English mathematician whose investigations into probability, published posthumously, led to what is now known as Bayes' theorem and a new branch of applications of probability theory. The theorem calculates the posterior probability of an event *A* given event *B* (that is, $P(A|B)$ on the basis of the prior probability of *A* (that is, $P(A)$). Basically, it tells us what sort of information can be retrodicted.

To introduce Bayes' theorem, let us return to our example. Jill wishes to know what action she should take but she lacks some essential information. She is underinformed. If she could learn what the probability is that the email was infected, given the fact that the email was placed in the quarantine folder, she could adopt the right course of action. Jill decides to run an ideal test on one million emails. The result is shown in Figure 18. The chance that an email might be infected before being blocked by the antivirus is 2%, but the chance that an email in the quarantine folder is

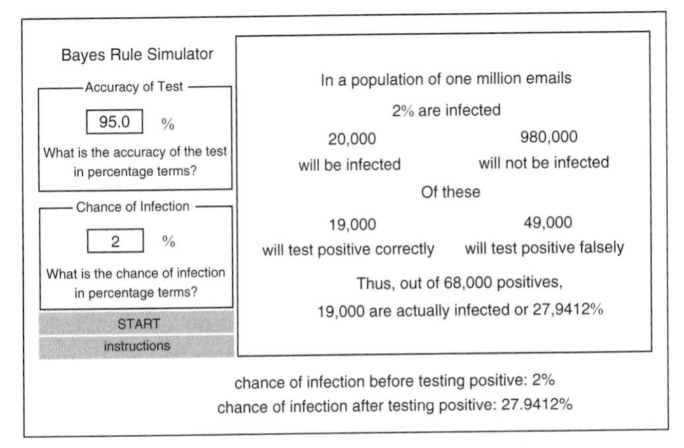


**18. A simple application of Bayes' theorem, adapted and reproduced with permission from Mikhael Shor, 'Bayes Rule Applet', *Game Theory.net*, (2005): http://www.gametheory.net/Mike/applets/Bayes/Bayes.html**



actually infected is roughly 28%. Jill has just acquired the relevant new information needed to shape her actions: clearly she should check her folder rather regularly.

Bayes' theorem formalizes Jill's reasoning in a well-known formula:

$$P(A/B) = \frac{P(B/A) \times P(A)}{P(B/A) \times P(A) + P(B/A^c) \times P(A^c)}$$

Let us now unpack what Bayes' theorem says. Jill is a smart girl. Maggie, a friend of hers, is not. She uses the same antivirus and receives roughly the same number of emails, with approximately the same quantity of infections but, when Jill explains to her that she should check her quarantine folder regularly, she is astonished. For she thought that, if the email was infected, then the antivirus blocked it, and since the quarantine folder contains only emails blocked by the antivirus, then all the emails in it must be infected. More formally, she reasoned that if A then B, and B is given, therefore A. Jill explains to Maggie that the previous inference is a typical logical mistake (a fallacy), but that she should not feel silly at all. For, consider Bayes' theorem once again. Look at the formula $P(B|A^c)$, where $A^c$ (the absolute complement) is just another notation for the negation of $A$. $P(B|A^c)$ indicates the probability that the antivirus blocks the email ($B$) when the email is not infected ($A^c$). Suppose we have perfect, infallible antivirus software. This will generate no false positives (no mistakes). But if there are no false positives, that is, if $P(B|A^c) = 0$, then $P(A|B) = 1$ and Bayes' theorem is degraded to a double implication: A if and only if B, and B is given, therefore A, which is not a fallacy and might be what Maggie had in mind. On the other hand, if there are some false positives, that is, if $P(B|A^c) > 0$, then $P(A|B) < 1$ and the formula bears a strong family resemblance to the fallacy in question, which is what Maggie might also have had in mind. Either way, Maggie was taking a shortcut (she disregarded the probabilities) to focus on the sort of information that she could extract from the fact that those emails were in the quarantine folder. And on the wise advice of being safe rather than sorry, she treated all its content as dangerous. The result is that Maggie is thrifty (she trusts many less items than Jill) by being logically greener (she relies on a reasoning that, although formally fallacious, can still be recycled to provide a quick and dirty way of extracting useful information from her environment). If this is unclear, the reader may try this last example.



John knows that, if the battery is flat, the engine will not start. Unfortunately, the engine does not start, so John calls the garage. The mechanic tells John that, given that fact that, if the battery is flat, the engine will not start, since the engine does not start, it must be the case that the battery is flat. John has learnt his Bayesian lesson from his daughter Jill, so he knows that the mechanic's reasoning is fallacious. But he also knows that it is fairly accurate, as a shortcut that gets things right most of the time: on average, engines that fail to start at all, making no noise and so forth, are the result of completely flat batteries. The mechanic took a shortcut, but he should have added 'probably'.

# Chapter 8
# The ethics of information

Our journey through the various concepts of information is almost complete. We started by looking at the information revolution, and we are now ready to see some of its ethical implications.

The previous chapters illustrated some crucial transformations brought about by ICTs in our lives. Moral life is a highly information-intensive game, so any technology that radically modifies the 'life of information' is bound to have profound moral implications for any moral player. Recall that we are talking about an ontological revolution, not just a change in communication technologies. ICTs, by radically transforming the context in which moral issues arise, not only add interesting new dimensions to old problems, but lead us to rethink, methodologically, the very grounds on which our ethical positions are based. Let us see how.

## Information ethics as a new environmental ethics

ICTs affect an agent's moral life in many ways. For the sake of simplicity, they can be schematically organized along three lines, in the following way. Suppose our moral agent *A* is interested in pursuing whatever she considers her best course of action, given her predicament. We shall assume that *A*'s evaluations and interactions have *some* moral value, but no specific value needs to be introduced at this stage. Intuitively, *A* can avail herself of some information (information as a *resource*) to generate some other information (information as a *product*) and, in so doing, affect her informational environment (information as *target*). This simple model, summarized in Figure 19, will help us to get some initial orientation in the multiplicity of issues belonging to information ethics (henceforth IE). I shall refer to it as the RPT model.

The RPT model is useful to rectify an excessive emphasis occasionally placed on specific technologies (this happens most notably in *computer* ethics), by highlighting the more fundamental phenomenon of information in all its varieties and long tradition. This was also Wiener's position and the various difficulties encountered in the conceptual foundations of computer ethics are arguably connected to the fact that the latter has not yet been recognized as primarily an environmental ethics, whose main concern should be the ecological management and wellbeing of the infosphere.



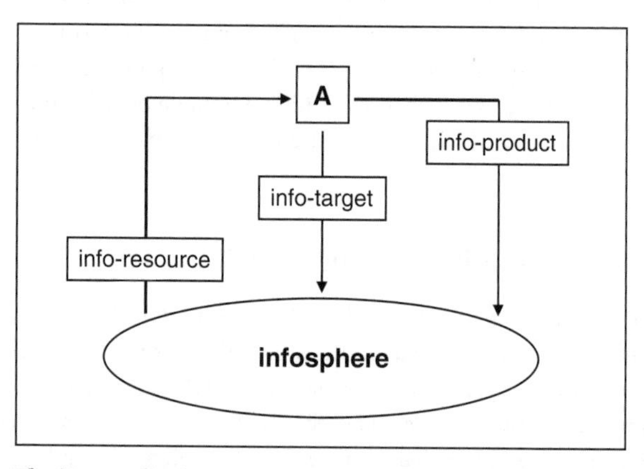


**19. The 'External' R(esource) P(roduct) T(arget) Model**

Since the appearance of the first works in the 1980s, information ethics has been claimed to be the study of moral issues arising from one or another of the three distinct 'information arrows' in the RPT model.

## Information-as-a-resource ethics

Consider first the crucial role played by information as a *resource* for *A*'s moral evaluations and actions. Moral evaluations and actions have an epistemic component, since *A* may be expected to proceed 'to the best of her information', that is, *A* may be expected to avail herself of whatever information she can muster, in order to reach (better) conclusions about what can and ought to be done in some given circumstances. Socrates already argued that a moral agent is naturally interested in gaining as much valuable information as the circumstances require, and that a well-informed agent is more likely to do the right thing. The ensuing 'ethical intellectualism' analyses evil and morally wrong behaviour as the outcome of deficient information. Conversely, *A*'s moral *responsibility* tends to be directly proportional to *A*'s degree of information: any decrease in the latter usually corresponds to a decrease in the former. This is the sense in which information occurs in the guise of judicial evidence. It is also the sense in which one speaks of *A*'s informed decision, informed consent, or well-informed participation. In Christian ethics, for example, even the worst sins can be forgiven in the light of the sinner's insufficient information, as a counterfactual evaluation is possible: had *A* been properly informed, *A* would have acted differently, and hence would not have sinned (Luke 23:34). In a secular context, Oedipus and Macbeth remind us how the mismanagement of informational resources may have tragic consequences.

From a 'resource' perspective, it seems that the moral machine needs information, and quite a lot of it, to function properly. However, even within the limited scope adopted by an analysis based solely on information as a resource, and hence a merely semantic view of the infosphere, care should be exercised, lest all ethical discourse is reduced to the nuances of higher quantity, quality, and intelligibility of informational resources. The more the better is not the only, nor always the best rule of thumb. For the (sometimes explicit and conscious) withdrawal of information can often make a significant difference. *A* may need to lack (or preclude herself from accessing) some information in order to achieve morally desirable goals, such as protecting anonymity, enhancing fair treatment, or implementing unbiased evaluation. Rawls' 'veil of ignorance' famously exploits precisely this aspect of information-as-a-resource, in order to develop an impartial approach to justice in terms of fairness. Being informed is not always a blessing and might even be morally wrong or dangerous.



Whether the (quantitative and qualitative) presence or the (total) absence of information-as-a-resource is in question, there is a perfectly reasonable sense in which information ethics may be described as the study of the moral issues arising from 'the triple A': *availability*, *accessibility*, and *accuracy* of informational resources, independently of their format, kind, and physical support. Examples of issues in information ethics understood as an information-as-resource ethics are the so-called *digital divide*, the problem of *infoglut*, and the analysis of the *reliability* and *trustworthiness* of information sources.

## Information-as-a-product ethics

A second, but closely related sense in which information plays an important moral role is as a *product* of *A*'s moral evaluations and actions. *A* is not only an information consumer but also an information producer, who may be subject to constraints while being able to take advantage of opportunities. Both constraints and opportunities call for an ethical analysis. Thus, information ethics, understood now as information-as-a-product ethics, may cover moral issues arising, for example, in the context of *accountability, liability, libel legislation, testimony, plagiarism, advertising, propaganda, misinformation*, and more generally the *pragmatic rules of communication*. The analysis of the immorality of *lying* offered by Immanuel Kant (1724–1804) is one of the best known case studies in the philosophical literature concerning this kind of information ethics. Cassandra and Laocoon, hopelessly warning the Trojans against the Greeks' wooden horse, remind us how the ineffective management of informational products may have tragic consequences.

## Information-as-a-target ethics

Independently of *A*'s information input (informational resources) and output (informational products), there is a third sense in which information may be subject to ethical analysis, namely when *A*'s moral evaluations and actions affect the informational environment. Examples include *A*'s respect for, or breach of, someone's information *privacy* or *confidentiality*. *Hacking*, understood as the unauthorized access to a (usually computerized) information system, is another good example. It is not uncommon to mistake it for a problem to be discussed within the conceptual frame of an ethics of informational resources. This misclassification allows the hacker to defend his position by arguing that no use (let alone misuse) of the accessed information has been made. Yet hacking, properly understood, is a form of breach of privacy. What is in question is not what *A* does with the information, which has been accessed without authorization, but what it means for an informational environment to be accessed by *A* without authorization. So the analysis of hacking belongs to an information-as-a-target ethics. Other issues here include *security, vandalism* (from the burning of libraries and books to the dissemination of viruses), *piracy, intellectual property, open source, freedom of expression, censorship, filtering*, and *contents control. Of the Liberty of*

*Thought and Discussion* by John Stuart Mill (1806–1873) is a classic of information ethics interpreted as information-as-target ethics. Juliet, simulating her death, and Hamlet, arranging for his father's homicide to be re-enacted, show how the unsafe management of one's informational environment may have tragic consequences.

## The limits of any microethical approach to information ethics

At the end of this overview, it seems that the RPT model may help one to get some initial orientation in the multiplicity of issues belonging to different interpretations of information ethics. Despite its advantages, however, the model can be criticized for being inadequate, in two respects.



On the one hand, the model is too simplistic. Arguably, several important issues belong *mainly but not only* to the analysis of just one 'informational arrow'. The reader may have already thought of several examples that illustrate the problem: someone's testimony is someone else's trustworthy information; *A*'s responsibility may be determined by the information *A* holds, but it may also concern the information *A* issues; censorship affects *A* both as a user and as a producer of information; disinformation (i.e. the deliberate production and distribution of false and misleading contents) is an ethical problem that concerns all three 'informational arrows'; freedom of speech also affects the availability of offensive content (e.g. child pornography, violent content, and socially, politically, or religiously disrespectful statements) that might be morally questionable and should not circulate.

On the other hand, the model is insufficiently inclusive. There are many important issues that cannot easily be placed on the map at all, for they really emerge from, or supervene on, the interactions among the 'informational arrows'. Two significant examples may suffice: 'big brother', that is, the problem of *monitoring and controlling* any information that might concern *A*; and the debate about information *ownership* (including copyright and patents legislation) and *fair use*, which affects both users and producers, while shaping their informational environment.

So the criticism is reasonable. The RPT model is indeed inadequate. Yet why it is inadequate and how it can be improved are different matters. The tripartite analysis just provided is unsatisfactory, despite its partial usefulness, precisely because any interpretation of information ethics based on only one of the 'informational arrows' is bound to be too reductive. As the examples mentioned above emphasize, supporters of narrowly constructed interpretations of information ethics as a *microethics* (that is a practical, field-dependent, applied, and professional ethics) are faced by the problem of being unable to cope with a large variety of relevant issues, which remain either uncovered or inexplicable. The model shows that idiosyncratic versions of information ethics, which privilege only some limited aspects of the *information cycle*, are unsatisfactory. We should not use the model to attempt to pigeonhole problems neatly, which is impossible. We should rather exploit it as a useful scheme to be superseded, in view of a more encompassing approach to information ethics as a *macroethics*, that is, a theoretical, field-independent, applicable ethics.



A more encompassing approach to information ethics needs to take three steps. It must bring together the three 'informational arrows'. It has to consider the whole information cycle. And it needs to take seriously the ontological shift in the nature of the infosphere emphasized in the first chapter. This means analysing informationally all entities involved (including the moral agent *A*) and their changes, actions, and interactions, and treating them not apart from, but as part of the informational environment to which they belong as informational systems themselves. Whereas the first two steps do not pose particular problems, and may be shared by other approaches to information ethics, the third step is crucial but involves an 'update' in the ontological conception of 'information' in question. Instead of limiting the analysis to (veridical) semantic contents – as any narrower interpretation of information ethics as a microethics inevitably does – an ecological approach to information ethics *also* treats information as an entity as well. In other words, we move from a broadly constructed epistemological or semantic conception of information ethics – in which information may be roughly equivalent to news or contents – to one which is typically ontological, and treats information as equivalent to patterns or entities in the world. Thus, in the revised RPT model, represented in Figure 20, the agent is embodied and embedded, as an informational agent or inforg, in an equally informational environment.

A simple analogy may help to introduce this new perspective. Imagine looking at the whole universe from a chemical perspective. Every entity and process will satisfy a certain chemical



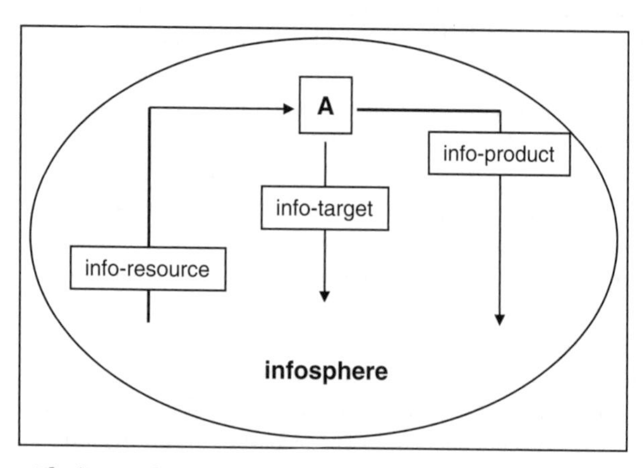


**20. The 'Internal' R(esource) P(roduct) T(arget) Model: the Agent A is correctly embedded within the infosphere**

description. A human being, for example, will be mostly water. Now consider an informational perspective. The same entities will be described as clusters of data, that is, as informational objects. More precisely, our agent *A* (like any other entity) will be a discrete, self-contained, encapsulated package containing (i) the appropriate data structures, which constitute the nature of the entity in question, that is, the state of the object, its unique identity, and its attributes; and (ii) a collection of operations, functions, or procedures, which are activated by various interactions or stimuli (that is, messages received from other objects or changes within itself), and correspondingly define how the object behaves or reacts to them. At this level of analysis, informational systems as such, rather than just living systems in general, are raised to the role of agents and patients of any action, with environmental processes, changes, and interactions equally described informationally.

Understanding the *nature* of information ethics ontologically rather than epistemologically modifies the interpretation of its *scope*. Not only can an ecological information ethics gain a global view of the whole life cycle of information, thus overcoming the limits of other microethical approaches, but it can also claim a role as a macroethics, that is, as an ethics that concerns the whole realm of reality, as explained in the next section.

## Information ethics as a macroethics

A straightforward way to introduce information ethics as a general approach to moral issues is by comparing it to environmental ethics. Environmental ethics grounds its analysis of the moral standing of biological entities and ecosystems on the intrinsic worthiness of *life*, and on the intrinsically negative value of *suffering*. It is biocentric. It seeks to develop a patient-oriented ethics in which the 'patient' may be not only a human being, but also any form of life. Indeed, land ethics extends the concept of patient to any component of the environment, thus coming close to the approach defended by information ethics. Any form of life is deemed to enjoy some essential proprieties or moral interests that deserve and demand to be respected, at least minimally and relatively, that is, in a possibly overridable sense, when contrasted to other interests. So biocentric ethics argues that the nature and wellbeing of the patient (the receiver) of any action constitute (at least partly) its moral standing and that the latter makes important claims on the interacting agent. These claims in principle ought to contribute to the guidance of the agent's ethical decisions and the constraint of the agent's moral behaviour. The 'receiver' of the action, the patient, is placed at the core of the ethical discourse, as a centre of moral concern, while the 'transmitter' of any moral action, the agent, is moved to its periphery.

Substitute now 'life' with 'existence' and it should become clear what information ethics amounts to. It is an ecological ethics that is still patient-oriented but replaces *biocentrism* with *ontocentrism*. It suggests that there is something even more elemental than life, namely *being* – that is, the existence and flourishing of all entities and their global environment – and something even more fundamental than suffering, namely *entropy*. The latter is most emphatically *not* the concept of thermodynamic entropy discussed in Chapter 5, the level of mixedupness of a system. Entropy here refers to any kind of *destruction*, *corruption*, *pollution*, and *depletion* of informational objects (mind, not just of information as semantic content), that is, any form of impoverishment of reality. Information ethics then provides a common vocabulary to understand the whole realm of *being* informationally. It holds that *being*/information has an intrinsic worthiness. It substantiates this position by recognizing that any informational entity has a right to persist in its own status, and a right to flourish, i.e. to improve and enrich its existence and essence. As a consequence of such 'rights', information ethics evaluates the duty of any moral agent in terms of contribution to the growth of the *infosphere* and any process, action, or event that negatively affects the whole infosphere – not just an informational entity – as an increase in its level of entropy and hence an instance of evil.

In information ethics, the ethical discourse concerns any entity, understood informationally, that is, not only all persons, their cultivation, wellbeing, and social interactions, not only animals, plants, and their proper natural life, but also anything that exists, from paintings and books to stars and stones; anything that may or will exist, like future generations; and anything that was but is no more, like our ancestors or old civilizations. Information ethics is impartial and universal because it brings to ultimate completion the process of enlargement of the concept of what may count as a centre of a (no matter how minimal) moral claim, which now includes every instance of *being* understood informationally, no matter whether physically implemented or not. In this respect, information ethics holds that every entity, as an expression of *being*, has a dignity, constituted by its mode of existence and essence (the collection of all the elementary proprieties that constitute it for what it is), which deserve to be respected (at least in a minimal and overridable sense), and hence place moral claims on the interacting agent and ought to contribute to the constraint and guidance of his ethical decisions and behaviour. This *ontological equality principle* means that any form of reality (any instance of information/*being*), simply for the fact of being what it is, enjoys a minimal, initial, overridable, equal right to exist and develop in a way which is appropriate to its nature. The conscious recognition of the ontological equality principle presupposes a disinterested judgement of the moral situation from an objective perspective, i.e. a perspective which is as non-anthropocentric as possible. Moral behaviour is less likely without this epistemic virtue. The application of the ontological equality principle is achieved, whenever actions are impartial, universal, and 'caring'. At the roots of this approach lies the *ontic trust* binding agents and patients. A straightforward way of clarifying the concept of ontic trust is by drawing an analogy with the concept of 'social contract'.

Various forms of contractualism (in ethics) and contractarianism (in political philosophy) argue that moral obligation, the duty of political obedience, or the justice of social institutions gain their support from a so-called 'social contract'. This may be a hypothetical agreement between the parties constituting a society (e.g. the people and the sovereign, the members of a community, or the individual and the state). The parties agree to accept the terms of the contract and thus obtain some rights in exchange for some freedoms that, allegedly, they would enjoy in a hypothetical state of nature. The rights and responsibilities of the parties subscribing to the agreement are the terms of the social contract, whereas the society, state, group, etc. is the entity created for the purpose of enforcing the agreement. Both rights and freedoms are not fixed and may vary, depending on the interpretation of the social contract.



Interpretations of the theory of the social contract tend to be highly (and often unknowingly) anthropocentric (the focus is only on human rational agents) and stress the coercive nature of the agreement. These two aspects are not characteristic of the concept of *ontic trust*, but the basic idea of a fundamental agreement between parties as a foundation of moral interactions is sensible. In the case of the ontic trust, it is transformed into a primeval, entirely hypothetical *pact*, logically predating the social contract, which all agents cannot but sign when they come into existence, and that is constantly renewed in successive generations.

In the English legal system, a trust is an entity in which someone (the trustee) holds and manages the former assets of a person (the trustor, or donor) for the benefit of certain persons or entities (the beneficiaries). Strictly speaking, nobody owns the assets, since the trustor has donated them, the trustee has only legal ownership and the beneficiary has only equitable ownership.

Now, the logical form of this sort of agreement can be used to model the ontic trust, in the following way:

- the assets or 'corpus' is represented by the world, including all existing agents and patients;
- the donors are all past and current *generations* of agents;
- the trustees are all current *individual* agents;
- the beneficiaries are all current and future *individual* agents and patients.

By coming into being, an agent is made possible thanks to the existence of other entities. It is therefore bound to all that already exists, both *unwillingly* and *inescapably*. It *should be* so also *caringly*. *Unwillingly*, because no agent wills itself into existence, though every agent can, in theory, will itself out of it. *Inescapably*, because the ontic bond may be broken by an agent only at the cost of ceasing to exist as an agent. Moral life does not begin with an act of freedom but it may end with one. *Caringly* because participation in reality by any entity, including an agent – that is, the fact that any entity is an expression of what exists – provides a right to existence and an invitation (not a duty) to respect and take care of other entities. The pact then involves no coercion, but a mutual relation of appreciation, gratitude, and care, which is fostered by the recognition of the dependence of all entities on each other. Existence begins with a gift, even if possibly an unwanted one. A foetus will be initially only a beneficiary of the world. Once she is born and has become a full moral agent, she will be, as an individual, both a beneficiary and a trustee of the world. She will be in charge of taking care of the world, and, insofar as she is a member of the generation of living agents, she will also be a donor of the world. Once dead, she will leave the world to other agents after her and thus become a member of the generation of donors. In short, the life of a human agent becomes a journey from being only a beneficiary to being only a donor, passing through the stage of being a responsible trustee of the world. We begin our career of moral agents as strangers to the world; we should end it as friends of the world.

The obligations and responsibilities imposed by the ontic trust will vary depending on circumstances but, fundamentally, the expectation is that actions will be taken or avoided in view of the welfare of the whole world.

The crucial importance of the radical change in ontological perspective cannot be overestimated. Bioethics and environmental ethics fail to achieve a level of complete impartiality, because they are still biased against what is inanimate, lifeless, intangible, or abstract (even land ethics is biased against technology and artefacts, for example). From their perspective, only what is intuitively alive deserves to be considered as a proper centre of moral claims, no matter how minimal, so a whole universe escapes their attention. Now, this is precisely the fundamental limit overcome by information ethics, which further lowers the minimal condition that needs to be satisfied, in order to qualify as a centre of moral concern, to the common factor shared by any entity, namely its informational state. And since any form of *being* is in any case also a coherent body of information, to say that information ethics is infocentric is tantamount to interpreting it, correctly, as an ontocentric theory.

The result is that all entities, *qua* informational objects, have an intrinsic moral value, although possibly quite minimal and overridable, and hence they can count as moral patients, subject to some equally minimal degree of moral respect understood as *a disinterested, appreciative, and careful attention*. As the philosopher Arne Naess (1912–2009) has maintained, 'all things in the biosphere have an equal right to live and blossom'. There seems to be no good reason not to adopt a higher and more inclusive, ontocentric perspective. Not only inanimate but also ideal, intangible, or intellectual objects can have a minimal degree of moral value, no matter how humble, and so be entitled to some respect.

There is a famous passage, in one of Albert Einstein's letters, that summarizes well this ontic perspective advocated by information ethics. A few years before his death, Einstein received a letter from a 19-year-old girl grieving over the loss of her younger sister. She wished to know whether the famous scientist might have something to say to comfort her. On 4 March 1950, Einstein replied to her:

> A human being is part of the whole, called by us 'universe', a part limited in time and space. He experiences himself, his thoughts and feelings, as something separated from the rest, a kind of optical delusion of his consciousness. This delusion is a kind of prison for us, restricting us to our personal desires and to affection for a few persons close to us. Our task must be to free ourselves from our prison by widening our circle of compassion to embrace all humanity and the whole of nature in its beauty. Nobody is capable of achieving this completely, but the striving for such achievement is in itself a part of the liberation and a foundation for inner security.

Deep ecologists have already argued that inanimate things too can have some intrinsic value. In a famous article, the historian Lynn Townsend White, Jr (1907–1987) asked:

> Do people have ethical obligations toward rocks? [and answered that] To almost all Americans, still saturated with ideas historically dominant in Christianity . . . the question makes no sense at all. If the time comes when to any considerable group of us such a question is no longer ridiculous, we may be on the verge of a change of value structures that will make possible measures to cope with the growing ecologic crisis. One hopes that there is enough time left.

According to information ethics, this is the right ecological perspective *and* it makes perfect sense for any religious or spiritual tradition (including the Judeo-Christian one) for which the whole universe is a divine creation, is inhabited by the divine, and is a gift to humanity, of which the latter needs to take care. Information ethics translates all this into informational terms. If something can be a moral patient, then its nature can be taken into consideration by a moral agent *A*, and contribute to shaping *A*'s action, no matter how minimally. In more metaphysical terms, information ethics argues that all aspects and instances of *being* are worth some initial, perhaps minimal and overridable, form of moral respect.

Enlarging the conception of what can count as a centre of moral respect has the advantage of enabling one to make sense of the innovative nature of ICTs, as providing a new and powerful conceptual frame. It also enables one to deal more satisfactorily with the original character of some of its moral issues, by approaching them from a theoretically strong perspective. Through time, ethics has steadily moved from a narrow to a more inclusive concept of what can count as a centre of moral worth, from the citizen to the biosphere. The emergence of the infosphere, as a new Athenian environment in which human beings spend much of their lives, explains the need to enlarge further the conception of what can qualify as a moral patient. Thus, information ethics represents the most recent development in this ecumenical trend, and an ecological approach without a biocentric bias. It translates environmental ethics in terms of infosphere and informational objects, for the space we inhabit is not just the earth.

# Epilogue: the marriage of *physis* and *techne*

It seems that, in view of the important change in our self-understanding (Chapter 1) and of the sort of ICT-mediated interactions that we will increasingly enjoy with other agents, whether biological or artificial (Chapter 8), the best way of tackling the new ethical challenges posed by ICTs may be from an environmental approach. This should not privilege the natural or untouched, but treat as authentic and genuine all forms of existence and behaviour, even those based on artificial, synthetic, or engineered artefacts. This sort of holistic *environmentalism* requires a change in our metaphysical perspective about the relationship between *physis* (nature, reality) and *techne* (practical science and its applications).

Whether *physis* and *techne* may be reconcilable is not a question that has a predetermined answer, waiting to be divined. It is more like a practical problem, whose feasible solution needs to be devised. With an analogy, we are not asking whether two chemicals could mix but rather whether a marriage may be successful. There is plenty of room for a positive answer, provided the right sort of commitment is made. It seems beyond doubt that a successful marriage between *physis* and *techne* is vital for our future and hence worth our sustained efforts. Information societies increasingly depend upon technology to thrive, but they equally need a healthy, natural environment to flourish. Try to imagine the world not tomorrow or next year, but next century, or next millennium: a divorce between *physis* and *techne* would be utterly disastrous both for our welfare and for the wellbeing of our habitat. This is something that technophiles and green fundamentalists must come to understand. Failing to negotiate a fruitful, symbiotic relationship between technology and nature is not an option.

Fortunately, a successful marriage between *physis* and *techne* is achievable. True, much more progress needs to be made. The physics of information can be highly energy-consuming and hence potentially unfriendly towards the environment. In 2000, data centres consumed 0.6% of the world's electricity. In 2005, the figure had increased to 1%. They are now responsible for more carbon dioxide emissions per year than Argentina or the Netherlands and, if current trends hold, their emissions will have grown four-fold by 2020, reaching 670 million tonnes. By then, it is estimated that the carbon footprint of ICTs will be higher than that of aviation. However, according to recent studies, ICTs will also help to eliminate almost 8 metric gigatons of greenhouse gas emissions annually by 2020, which is equivalent to 15% of global emissions today and five times more than the estimated emissions from ICTs in 2020. This positive and improvable balance leads me to a final comment.

The greenest machine is a machine with 100% energy efficiency. Unfortunately, this is equivalent to a perpetual motion machine and we saw in Chapter 5 that the latter is simply a pipe dream. However, we also know that such an impossible target can be increasingly approximated: energy waste can be dramatically reduced and energy efficiency can be highly increased (the two processes are not necessarily the same: compare recycling versus doing more with less). Often, both kinds of processes may be fostered only by relying on significant improvements in the management of information (e.g. to build and run hardware and processes better). So here is how we may reinterpret

Socrates' ethical intellectualism, encountered in the previous chapter: we do evil because we do not know better, in the sense that the better the information management is, the less moral evil is caused. With a proviso, though: some ethical theories seem to assume that the moral game, played by agents in their environments, may be won absolutely, i.e. not in terms of higher scores, but by scoring perhaps very little as long as no moral loss or error occurs, a bit like winning a football game by scoring only one goal as long as none is received. It seems that this absolute view has led different parties to underestimate the importance of successful compromises. Imagine an environmentalist unable to accept any technology responsible for some level of carbon dioxide emission, no matter how it may counterbalance it. The more realistic and challenging view is that moral evil is unavoidable, so that the real effort lies in limiting it and counterbalancing it with more moral goodness.

ICTs can help us in our fight against the destruction, impoverishment, vandalism, and waste of both natural and human resources, including historical and cultural ones. So they can be a precious ally in what I have called elsewhere *synthetic environmentalism* or *e-nvironmentalism*. We should resist any Greek epistemological tendency to treat *techne* as the Cinderella of knowledge; any absolutist inclination to accept no moral balancing between some unavoidable evil and more goodness; or any modern, reactionary, metaphysical temptation to drive a wedge between naturalism and constructionism, by privileging the former as the only authentic dimension of human life. The challenge is to reconcile our roles as informational organisms and agents within nature and as stewards of nature. The good news is that it is a challenge we can meet. The odd thing is that we are slowly coming to realize that we have such a hybrid nature. The turning point in this process of self-understanding is what I have defined in Chapter 1 as the *fourth revolution*.

# References

## Introduction

W. Weaver, 'The Mathematics of Communication', *Scientific American*, 1949, 181(1), 11–15.

C. E. Shannon, *Collected Papers*, edited by N. J. A. Sloane and A. D. Wyner (New York: IEEE Press, 1993).

C. E. Shannon and W. Weaver, *The Mathematical Theory of Communication* (Urbana, IL: University of Illinois Press, 1949; reprinted 1998).

## Chapter 1

L. Floridi, 'A Look into the Future Impact of ICT on Our Lives', *The Information Society*, 2007, 23(1), 59–64.

S. Freud, 'A Difficulty in the Path of Psycho-Analysis', *The Standard Edition of the Complete Psychological Works of Sigmund Freud*, XVII (London: Hogarth Press, 1917–19), 135–44.

## Chapter 2

J. Barwise and J. Seligman, *Information Flow: The Logic of Distributed Systems* (Cambridge: Cambridge University Press, 1997).

G. Bateson, *Steps to an Ecology of Mind* (Frogmore, St Albans: Paladin, 1973).

T. M. Cover and J. A. Thomas, *Elements of Information Theory* (New York; Chichester: Wiley, 1991).

F. I. Dretske, *Knowledge and the Flow of Information* (Oxford: Blackwell, 1981).
D. S. Jones, *Elementary Information Theory* (Oxford: Clarendon Press, 1979).
D. M. MacKay, *Information, Mechanism and Meaning* (Cambridge, MA: MIT Press, 1969).
J. R. Pierce, *An Introduction to Information Theory: Symbols, Signals and Noise*, 2nd edn (New York: Dover Publications, 1980).
A. M. Turing, 'Computing Machinery and Intelligence', *Minds and Machines*, 1950, 59, 433–60.

## Chapter 3

C. Cherry, *On Human Communication: A Review, a Survey, and a Criticism*, 3rd edn (Cambridge, MA; London: MIT Press, 1978).
A. Golan, 'Information and Entropy Econometrics – Editor's View', *Journal of Econometrics*, 2002, 107(1–2), 1–15.
P. C. Mabon, *Mission Communications: The Story of Bell Laboratories* (Murray Hill, NJ: Bell Telephone Laboratories, 1975).
C. E. Shannon and W. Weaver, *The Mathematical Theory of Communication* (Urbana, IL: University of Illinois Press, 1949; reprinted 1998).



## Chapter 4

F. I. Dretske, *Knowledge and the Flow of Information* (Oxford: Blackwell, 1981).
J. Barwise and J. Seligman, *Information Flow: The Logic of Distributed Systems* (Cambridge: Cambridge University Press, 1997).
Y. Bar-Hillel, *Language and Information: Selected Essays on Their Theory and Application* (Reading, MA; London: Addison-Wesley, 1964).
M. D'Agostino and L. Floridi, 'The Enduring Scandal of Deduction: Is Propositional Logic Really Uninformative?', *Synthese*, 2009, 167(2), 271–315.
L. Floridi, 'Outline of a Theory of Strongly Semantic Information', *Minds and Machines*, 2004, 14(2), 197–222.
J. Hintikka, *Logic, Language-Games and Information: Kantian Themes in the Philosophy of Logic* (Oxford: Clarendon Press, 1973).

K. R. Popper, *Logik Der Forschung: Zur Erkenntnistheorie Der Modernen Naturwissenschaft* (Wien: J. Springer, 1935).

## Chapter 5

P. Ball, 'Universe Is a Computer', *Nature News*, 3 June 2002.
C. H. Bennett, 'Logical Reversibility of Computation', *IBM Journal of Research and Development*, 1973, 17(6), 525–32.
R. Landauer, 'Irreversibility and Heat Generation in the Computing Process', *IBM Journal of Research and Development*, 1961, 5(3), 183–91.
S. Lloyd, 'Computational Capacity of the Universe', *Physical Review Letters*, 2002, 88(23), 237901–4.
J. C. Maxwell, *Theory of Heat* (Westport, CT: Greenwood Press, 1871).
J. A. Wheeler, 'Information, Physics, Quantum: The Search for Links', in *Complexity, Entropy, and the Physics of Information*, edited by W. H. Zureck (Redwood City, CA: Addison Wesley, 1990).
N. Wiener, *Cybernetics or Control and Communication in the Animal and the Machine*, 2nd edn (Cambridge, MA: MIT Press, 1961).



## Chapter 6

E. Schrödinger, *What Is Life? The Physical Aspect of the Living Cell* (Cambridge: Cambridge University Press, 1944).

## Chapter 7

M. D. Davis and O. Morgenstern, *Game Theory: A Nontechnical Introduction* (London: Dover Publications, 1997).
J. Nash, 'Non-Cooperative Games', *Annals of Mathematics*, Second Series, 1951, 54(2), 286–95.

## Chapter 8

A. Einstein, *Ideas and Opinions* (New York: Crown Publishers, 1954).
A. Naess, 'The Shallow and the Deep, Long-Range Ecology Movement', *Inquiry*, 1973, 16, 95–100.

J. Rawls, *A Theory of Justice*, revised edn (Oxford: Oxford University Press, 1999).
L. J. White, 'The Historical Roots of Our Ecological Crisis', *Science*, 1967, 155, 1203–7.
N. Wiener, *The Human Use of Human Beings: Cybernetics and Society*, revised edn (Boston, MA: Houghton Mifflin, 1954).

# Index

# F



# G



# H



# I

## Expand your collection of
# VERY SHORT INTRODUCTIONS

1. Classics
2. Music
3. Buddhism
4. Literary Theory
5. Hinduism
6. Psychology
7. Islam
8. Politics
9. Theology
10. Archaeology
11. Judaism
12. Sociology
13. The Koran
14. The Bible
15. Social and Cultural Anthropology
16. History
17. Roman Britain
18. The Anglo-Saxon Age
19. Medieval Britain
20. The Tudors
21. Stuart Britain
22. Eighteenth-Century Britain
23. Nineteenth-Century Britain
24. Twentieth-Century Britain
25. Heidegger
26. Ancient Philosophy
27. Socrates
28. Marx
29. Logic
30. Descartes
31. Machiavelli
32. Aristotle
33. Hume
34. Nietzsche
35. Darwin
36. The European Union
37. Gandhi
38. Augustine
39. Intelligence
40. Jung
41. Buddha
42. Paul
43. Continental Philosophy
44. Galileo
45. Freud
46. Wittgenstein
47. Indian Philosophy
48. Rousseau
49. Hegel
50. Kant
51. Cosmology
52. Drugs
53. Russian Literature
54. The French Revolution
55. Philosophy
56. Barthes
57. Animal Rights
58. Kierkegaard
59. Russell
60. Shakespeare
61. Clausewitz
62. Schopenhauer
63. The Russian Revolution
64. Hobbes

65. World Music
66. Mathematics
67. Philosophy of Science
68. Cryptography
69. Quantum Theory
70. Spinoza
71. Choice Theory
72. Architecture
73. Poststructuralism
74. Postmodernism
75. Democracy
76. Empire
77. Fascism
78. Terrorism
79. Plato
80. Ethics
81. Emotion
82. Northern Ireland
83. Art Theory
84. Locke
85. Modern Ireland
86. Globalization
87. The Cold War
88. The History of Astronomy
89. Schizophrenia
90. The Earth
91. Engels
92. British Politics
93. Linguistics
94. The Celts
95. Ideology
96. Prehistory
97. Political Philosophy
98. Postcolonialism
99. Atheism
100. Evolution
101. Molecules
102. Art History
103. Presocratic Philosophy
104. The Elements
105. Dada and Surrealism
106. Egyptian Myth
107. Christian Art
108. Capitalism
109. Particle Physics
110. Free Will
111. Myth
112. Ancient Egypt
113. Hieroglyphs
114. Medical Ethics
115. Kafka
116. Anarchism
117. Ancient Warfare
118. Global Warming
119. Christianity
120. Modern Art
121. Consciousness
122. Foucault
123. The Spanish Civil War
124. The Marquis de Sade
125. Habermas
126. Socialism
127. Dreaming
128. Dinosaurs
129. Renaissance Art
130. Buddhist Ethics
131. Tragedy
132. Sikhism
133. The History of Time
134. Nationalism
135. The World Trade Organization
136. Design
137. The Vikings
138. Fossils
139. Journalism
140. The Crusades

141. Feminism
142. Human Evolution
143. The Dead Sea Scrolls
144. The Brain
145. Global Catastrophes
146. Contemporary Art
147. Philosophy of Law
148. The Renaissance
149. Anglicanism
150. The Roman Empire
151. Photography
152. Psychiatry
153. Existentialism
154. The First World War
155. Fundamentalism
156. Economics
157. International Migration
158. Newton
159. Chaos
160. African History
161. Racism
162. Kabbalah
163. Human Rights
164. International Relations
165. The American Presidency
166. The Great Depression and The New Deal
167. Classical Mythology
168. The New Testament as Literature
169. American Political Parties and Elections
170. Bestsellers
171. Geopolitics
172. Antisemitism
173. Game Theory
174. HIV/AIDS
175. Documentary Film
176. Modern China
177. The Quakers
178. German Literature
179. Nuclear Weapons
180. Law
181. The Old Testament
182. Galaxies
183. Mormonism
184. Religion in America
185. Geography
186. The Meaning of Life
187. Sexuality
188. Nelson Mandela
189. Science and Religion
190. Relativity
191. History of Medicine
192. Citizenship
193. The History of Life
194. Memory
195. Autism
196. Statistics
197. Scotland
198. Catholicism
199. The United Nations
200. Free Speech
201. The Apocryphal Gospels
202. Modern Japan
203. Lincoln
204. Superconductivity
205. Nothing
206. Biography
207. The Soviet Union
208. Writing and Script
209. Communism

210. Fashion
211. Forensic Science
212. Puritanism
213. The Reformation
214. Thomas Aquinas
215. Deserts
216. The Norman Conquest
217. Biblical Archaeology
218. The Reagan Revolution
219. The Book of Mormon
220. Islamic History
221. Privacy
222. Neoliberalism
223. Progressivism
224. Epidemiology
225. Information

# INTELLIGENCE

## A Very Short Introduction

Ian J. Deary

Ian J. Deary takes readers with no knowledge about the science of human intelligence to a stage where they can make informed judgements about some of the key questions about human mental activities. He discusses different types of intelligence, and what we know about how genes and the environment combine to cause these differences; he addresses their biological basis, and whether intelligence declines or increases as we grow older. He charts the discoveries that psychologists have made about how and why we vary in important aspects of our thinking powers.

'There has been no short, up to date and accurate book on the science of intelligence for many years now. This is that missing book. Deary's informal, story-telling style will engage readers, but it does not in any way compromise the scientific seriousness of the book . . . excellent.'

**Linda Gottfredson, University of Delaware**

'Ian Deary is a world-class leader in research on intelligence and he has written a world-class introduction to the field . . . This is a marvellous introduction to an exciting area of research.'

**Robert Plomin, University of London**

www.oup.co.uk/isbn/0-19-289321-1

# ANARCHISM

## A Very Short Introduction

Colin Ward

The word 'anarchism' tends to conjure up images of aggressive protest against government. But is anarchism inevitably linked with violent disorder? Do anarchists adhere to a coherent ideology? What exactly is anarchism?

In this Very Short Introduction, Colin Ward considers anarchism from a variety of perspectives: theoretical, historical, and international, and by exploring key anarchist thinkers from Kropotkin to Chomsky. Among the questions he ponders are: can anarchy ever function effectively as a political force? Is it more 'organized' and 'reasonable' than is currently perceived? Whatever the politics of the reader, Ward's argument ensures that anarchism will be much better understood after reading this book.

'excellent introduction' – **The Guardian**

http://www.oup.co.uk/isbn/0-19-280477-4